THE BEHAVIOR AND
ECOLOGY OF THE AFRICAN BUFFALO

T0296925

THE BEHAVIOR AND ECOLOGY OF THE AFRICAN BUFFALO

MARK J. MLOSZEWSKI

CAMBRIDGE UNIVERSITY PRESS

Cambridge
London   New York   New Rochelle
Melbourne   Sydney

CAMBRIDGE UNIVERSITY PRESS
Cambridge, New York, Melbourne, Madrid, Cape Town, Singapore,
São Paulo, Delhi, Dubai, Tokyo, Mexico City

Cambridge University Press
The Edinburgh Building, Cambridge CB2 8RU, UK

Published in the United States of America by Cambridge University Press, New York

www.cambridge.org
Information on this title: www.cambridge.org/9780521144681

© Cambridge University Press 1983

This publication is in copyright. Subject to statutory exception
and to the provisions of relevant collective licensing agreements,
no reproduction of any part may take place without the written
permission of Cambridge University Press.

First published 1983
First paperback printing 2010

A catalogue record for this publication is available from the British Library

Library of Congress Cataloguing in Publication data
Mloszewski, Mark J.
The behavior and ecology of the African
buffalo.
Bibliography: p.
Includes index.
1. African buffalo – Behavior. 2. Social
behavior in animals. I. Title.
QL737.U53M56   599.73′58   82-1153

ISBN 978-0-521-24478-7 Hardback
ISBN 978-0-521-14468-1 Paperback

Cambridge University Press has no responsibility for the persistence or
accuracy of URLs for external or third-party Internet Web sites referred to in
this publication, and does not guarantee that any content on such Web sites is,
or will remain, accurate or appropriate.

Contents

Preface

When I compare my own field data on the African buffalo with various studies of domestic ungulate behavior, I feel dissatisfied. Domestic ungulate studies display abundant statistical data, based, in most cases, on hundreds of observations made under carefully controlled conditions; my buffalo data are often based on smaller samples and more variable conditions.

Then I recall the difficulties encountered in following wild ungulates on the move, the many days and nights required to obtain even a modest amount of data. I also realize that the opportunities to obtain more data in the future are diminishing rapidly, as habitats become significantly altered and populations are steadily destroyed – and this inclines me to present what I have obtained without further delay.

This study reports on the individual and collective behavior of the African buffalo. Field observations were made chiefly in parts of East and Central Africa, but also in West Africa. The time spent on field observations in Kenya, Zambia, Tanzania, and Zimbabwe (then Rhodesia) was 3,890 hours, from October 1964 to January 1978. No time log was kept in West Africa, as at that time – that is, previous to 1964 – my interest in the buffalo was only amateur and there was no commitment to a major study. Observations were also made on captive buffaloes in the zoological parks at Antwerp, Barcelona, Whipsnade, and elsewhere. The study was carried out on a part-time basis until 1977, a year devoted to it entirely.

The aims of this study were to learn, under wilderness conditions, about the social organization of the African buffalo, including the behavior and composition of large groups, interactions between pairs or small numbers of individuals, interspecific encounters, and solitary behavior.

It was obvious that alimentary, agonistic, and reproductive patterns would have to be recorded, and it was hoped that something could be learned about communications. The methods employed in the course of this work are discussed in Chapter 3.

I wish to acknowledge indebtedness and give my thanks to the National Parks and Wildlife Service of Zambia, the Kenya Game Department, the Tanzania National Parks, and the Department of National Parks and Wild Life Management of Rhodesia (now Zimbabwe), for their cooperation at various times during this study. In some cases the acknowledgments also refer to the predecessors or successors of the said departments, as names have sometimes been altered in the course of governmental modifications. I wish to give my grateful thanks to the Anglo-American Corporation, Ltd., for their partial but substantial material support of the Zambian part of the project. My thanks go also to the other organizations and persons who have in some measure contributed materially to the project, notably Zambia Safaris, Ltd., whose accommodation facilities and hospitality at their Lusaka headquarters were generously open to me at all times during the last year of the project. Thanks are extended to the Société Royale de Zoologie d'Anvers, Belgium, for its kind cooperation in allowing me to carry out a part of this study in its zoological gardens, and to the Regent's Park, London, and Whipsnade Zoos in England for useful information. During the literature search, the Mammal Section of the British Museum (Natural History), London, was most cooperative in letting me use their facilities for several weeks, and I must give special thanks to the Cambridge University Library for letting me in at a time when they were closed for annual inventory.

There are many individuals to whom warm personal thanks are due. The following persons have made contributions relevant to this work, some by a few remarks during a single conversation, others by donating an extraordinarily large part of their time. Some contributions were in the form of facilities or services that were valuable, sometimes indispensable, to the carrying out of this work. Hospitality was often included. I wish especially to thank Dr. C. S. Churcher (University of Toronto Zoology Department) who, besides contributing directly to Chapter 2, was the main critic of my manuscript. I wish to thank the following, in alphabetical order: Mr. W. F. H. Ansell, Mr. R. I. G. Attwell, Mr. W. Van den Bergh, Mr. M. C. Bromwich, Mr. Norman Carr, Mr. N. H. Chabwela, Game Guard S. Chola, Mr. R. A. Conant, Lt. Col. R. A. Critchley, Dr. A. W. Gentry, Dr. J. J. R. Grimsdell, Mr. J. E. Hazam, Game Scout Hlupa, Dr. A. Keast, Mr. Kula s/o Masila, Dr. H. S. Logsdon, Game Scout Sgt. Machapora, Game Scout Sgt. Mathalavana, Game Guard A. Mumba, Mr. T. Orford, Mr. J. Posselt, Mr. Van Puijenbroeck, Miss M. L. Richardson, Mr. Paul Russell, Dr. J. Sabater Pi, Mr. Y. Savidge, Mr.

E. Sicheepa, Miss M. P. E. Tan, the late Maj. E. Temple-Boreham, Game Scout Tomu, and Mr. C. G. N. Zyambo.

Mrs. Patrick Maguire redrafted parts b to e of Figure 1.1. Miss M. L. Richardson (now Mrs. Geoffrey Jenkins) drafted Figures 6.9, 6.10, 7.2, and 11.1 and contributed to Figures 4.1 and 5.2. Mrs. Alan Wardle helped in the early stages of manuscript typing. Drs. J. C. Barlow and R. D. James (Royal Ontario Museum) made available equipment and help in the production of sonograms. Mr. J. N. Glover (Photographic Section, Zoology Department, University of Toronto) applied his professional skill to produce photographic prints from color transparencies, some of which were on previously damaged film.

It is impossible for me to record all the instances of specialist, material, or moral support received by me in the course of this study, and therefore I extend my heartfelt thanks collectively to all who have helped. I am personally responsible for any shortcomings that might occur in the following text.

Mark J. Mloszewski

1

Introduction

Literature

Early literature

In 1553 the French physician-naturalist Pierre Belon (Bellonius) described an African bovid from Asamia, possibly in present-day Morocco, that Brooke (1873) suggests may have been a buffalo. Belon called this animal a little ox. Its horns, however, are described as ringed like those of gazelles and raised upon the frontal bones. Could it have been, not an "ox," but an antelope such as a hartebeest (*Alcelaphus buselaphus*) or a related type, for example, *Damaliscus* sp.? We cannot be sure, as no specimen remains, and furthermore Belon may have obtained the geographic locality only by hearsay (Figure 1.1).

There are a few mentions of the buffalo in the writings of seventeenth- and eighteenth-century Dutch settlers in southern Africa, for example, in van der Stel's narrative of his journey to Namaqualand (English trans. Waterhouse, 1932). Such descriptions contain little information other than indications of the buffalo's former range. Until about the middle of the nineteenth century there is only a trickle of new information on the African buffalo, mostly of a superficial nature. It consists mainly of notes relevant to taxonomy (e.g., Sparrman, 1779, 1785; Thunberg, 1811; Smith, 1827, 1850; Smuts, 1832; Smith, 1834; Gray, 1837, 1843; Hodgson, 1847) and of hunters' accounts (e.g., Cornwallis Harris, 1840).

In the second half of the nineteenth century the trickle increases substantially. For example, the subject index of the proceedings of the Zoological Society of London for 1830–60 has only one entry dealing with the African buffalo, but in the following 30 years there are over 60 such entries. Most of these deal either with the acquisition of live and dead specimens or with taxonomy. Papers that appeared during the

1

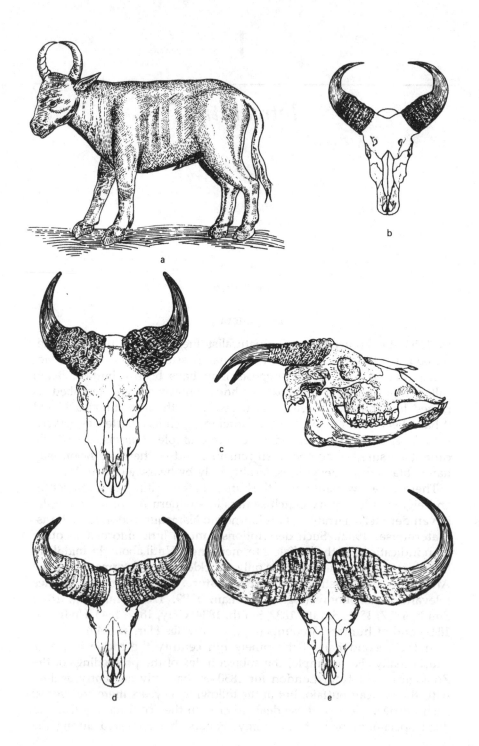

a

b

c

d

e

same period in the French bulletins of the Société Philomatique de Paris and of the Société Nationale d'Acclimatisation, in the German *Zoologisches Jahrbuch*, and in the notes of the Leyden Museum and others also deal mainly with points of taxonomic interest. Among the authors are Sclater (1864), Kirk (1864), Blyth (1866), Gray (1872), Brooke (1873, 1875), Johnson (1885), Rochebrune (1885), Cincinn (1887), Schlegel (1880), and Pechuel-Loesche (1888).

Some comment on the distribution of the African buffalo also begins in the second half of the nineteenth century (e.g., Temminck, 1853; Buckley, 1876).

Occasionally, useful accounts of the buffalo occur in explorers' and hunters' narratives of that period, for example, Livingstone (1857, 1865), Holub (1881), Selous (1881), and Bryden (1899). Most of the early notes on African buffalo behavior and social structure that may merit attention occur in such accounts. The value of these sources is very unequal. Some of the notes, especially those by Selous, simply report direct observations of buffaloes and contain a fair amount of detail, but some others appear overinterpreted or too fragmental. A few of these accounts, for example, von der Decken's *Travels in East Africa* (1869), have been edited to a damaging extent by persons without direct experience of the subject matter. Lydekker (1898, 1904, 1906, 1910, 1913, 1926) may have done most to launch the modern trend in African buffalo systematics.

Twentieth-century literature

African buffalo systematics were discussed and revised during and after Lydekker's time by a number of other authors. Contributions in

Figure 1.1 Some historic buffaloes: (a) This may be the first picture of an African buffalo to appear in a natural history text. It accompanies the description by Pierre Belon, a sixteenth-century French physician-naturalist, of a small buffalo claimed (with questionable accuracy) to have come from northern Africa. The illustrator may have seen an authentic *Syncerus* hide, to judge by the deep wrinkles on the neck and an indication of hairy tufting in the ear. The horns, if drawn from a specimen, may have belonged either to a *Syncerus caffer* of the *brachyceros* or *nanus* variety, or to an alcelaphine antelope, although the body is clearly meant to appear cattlelike. The depicted skulls, redrawn from Brooke (1873), were used in the early studies of the species by Sir Victor Brooke and others. (b) A female skull brought by a Captain Clapperton from central Africa in the 1830s (greatest horn spread about 51 cm); (c) two aspects of a male skull obtained by Brooke from E. Gerrard in 1872 (greatest horn spread about 69 cm); (d) a male skull brought by Sir Samuel Baker from Sudan or Ethiopia in the 1860s (greatest horn spread about 64 cm); (e) a male *S. c. caffer* skull represented in Brooke's paper and described as a "very fine specimen of the horns of *Bubalus caffer*" (Brooke 1873, p. 481) (greatest horn spread about 104 cm). It is now known that the span of such horns can slightly exceed 142 cm.

this field were made by Schwartz (1920), Christy (1929), Malbrant (1936), Schouteden (1945), Blancou (1954), Dalimier (1955), and Grubb (1971). Grubb, besides expressing his own views on African buffalo speciation, contributes a good review of the literature on this subject. Comments on Grubb's paper can be found in Groves (1975). A substantial paleontological contribution to the field of African buffalo speciation is found in Gentry (1979).

There exist a few brief studies, notes, and communications, mainly in French, English, and German, which have a local character and usually deal with more than one aspect of the buffalo. They may yield useful bits of information, depending on the reader's special interest, and some of them are cited in various parts of this work. Among those cited the most valuable are a paper by Pienaar (1969a) concerning the buffaloes of the Kruger Park, South Africa, and a paper by Henshaw and Greeling (1973) that describes the buffaloes of Yankeri Game Reserve, northeastern Nigeria, and reviews some relevant literature. Local studies that are not directly mentioned elsewhere in the present work but must not be overlooked include papers by Leuthold (1972) on a buffalo herd in the Tsavo Park, Kenya; Grimwood et al. (1958) on Zambian buffaloes; Baudenon (1952) on Togo buffaloes; and Monard (1935) on Angolan buffaloes.

The largest recent contribution is Sinclair's (1977) study of the buffaloes of the Serengeti region, northern Tanzania. This work contains a long list of references. There is a substantial body of literature dealing with parasitological and pathological conditions affecting the African buffalo, and much of this is listed in Sinclair (1977).

Finally, much varied information is contained in the annual and other reports of several African government game or wildlife departments. These sources, however, may be difficult to reach and tedious to search.

Biogeography

General notes

The main study areas are located on or near the great high plateau that covers roughly one-third of the African continent (Figure 1.2). This plateau occupies most of Africa south of the equator and in the east extends far into the northern hemisphere. The plateau with its margins probably comprises the entire geographic range of the largest races of the African buffalo, for example, *capensis* and *radicliffei* (the largest buffaloes found outside this region are of the slightly smaller *aequinoctialis* type).

This great highland is topographically diversified and includes flat to

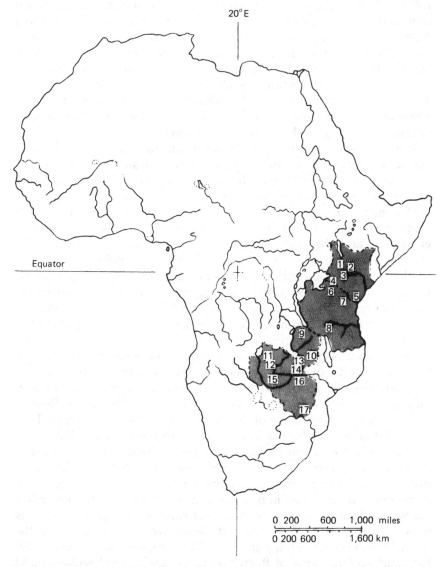

Figure 1.2. Study area. Shading indicates in general where most of the author's own observations were made. Localities considered particularly useful to the study are indicated. These do not always coincide with the largest remaining buffalo concentrations. Key: (1) Marallal–Baragoi; (2) Archer's Post–Samburu Game Reserve; (3) Mt. Kenya–Abardares; (4) Narok District (Lolgorien, Loita); (5) Coastal hinterland–Tsavo National Park; (6) Serengeti N.P.; (7) Ngurdoto Crater; (8) Ruaha N.P.; (9) Mporokoso–Sumbu; (10) Luangwa Valley; (11) Lukwakwa–W. Lunga N.P.; (12) Busanga and adjacent area; (13) Mkushi–Muchinga escarpment; (14) Zambezi Valley, north bank; (15) Nanzhila; (16) Zambezi Valley, south bank; (17) Gona-Re-Zhou G.R. (Mabalauta).

rolling plains, hilly areas, a few mountain chains (e.g., the Ethiopian mountains, the Ruwenzori, and the Drakensberg), and some isolated high mountains (e.g., Kilimanjaro, Kenya, and Elgon). There are also present very large elongated tectonic depressions known as rift valleys. Most of the high plateau is between 1,000 and 2,000 meters above sea level (a.s.l.), but on mountains elevations attain up to 5,895 meters a.s.l. (Kilimanjaro). Extensive floodplains occur along many larger rivers and around some lakes. Around the plateau margins occur large embayments of ground well below 1,000 meters a.s.l.

The surface rocks are largely Precambrian metamorphics and Paleozoic and Quaternary sedimentary formations. Mesozoic rocks also outcrop, but over smaller areas. Intrusive complexes and extrusive volcanic rocks are exposed in many places.

The main soil types of the great African high plateau are, in the north and south, the chestnut and brown soils typical of semiarid grasslands and the red to yellow savanna/rain forest ferrisols in the middle portion. On this background occur a few patches of dark gray to black subtropical vertisols. An area of montane soils is found in the Ethiopian part of the highland. A large expanse of red and gray desert soils occurs in the far southwest of the highland plateau, over the Kalahari area. The major soil types do not demarcate the plateau but extend into the neighboring lowlands.

The plateau is drained by several large river systems. The Orange, Cunene, and Cuanza systems empty directly into the Atlantic. The Limpopo and the great Zambezi system flow in the opposite direction into the Indian Ocean. Also important are two northwesterly flowing major tributary subsystems of the Congo river – the Kasai and the Lualaba – and so is the drainage northward into the Nile. The Ethiopian, Kenyan, and Tanzanian highlands are partly drained by several major streams which empty into the Indian Ocean – the Shaballe, Juba, Tana, Ruaha, and Ruvuma rivers. A large number of seasonal and some permanent streams flow into all these major rivers, which also collect the water of numerous surface and near-surface seepage lines, such as those called *dambos* in Zambia and Zimbabwe, and swamps. All the great African lakes, with the exception of Lake Chad, are located in this highland region, some of them occupying low ground within it and some occurring at relatively high elevations.

Mean annual rainfall over the high plateau varies between about 200 millimeters and a little under 2,000 millimeters. In the savanna areas it is of the order of 600 to 800 millimeters. The mean monthly temperatures vary between about 8 and 30°C. The most common temperatures are around 20°C.

The high plateau is covered mainly by savanna and arid land vegetation in its northern and southern parts and by dry deciduous woodland and savanna woodland in its middle portion. Patches of montane forest and meadow occur locally, particularly in a few localities between northern Ethiopia and the northern end of Lake Malawi. A few small patches of rain forest occur, for example, the Kakamega forest in western Kenya, the montane rain forests on Mount Kenya and the Ruwenzori range, and the Tsitsikama forest in the Cape Province of South Africa. Arid-land vegetation occurs in much of northern Kenya, in the Kalahari area, and elsewhere.

The savanna and deciduous woodland areas – the most common vegetation assemblages on the plateau – typically have one or two closely spaced rainy seasons annually. These coincide with the growth of grasses and are slightly preceded by the appearance of new leaves on deciduous trees. A large proportion of the year is rainless, when the grasses stop growing and trees are mainly without leaves.

The savannas support a large variety of grasses, typically of tufted growth habit (see also Chapter 6). Except for relatively rare areas of pure grassland savanna, a tree cover occurs which varies in density between types where individual trees are several tens of meters apart and types where crowns of trees nearly touch. The latter types grade into woodlands. Many species of trees are found in various parts of the high plateau, largely belonging to the genera *Acacia, Brachystegia, Julbernardia, Isoberlinia, Terminalia,* and *Albizia,* though many others also occur. *Colophospermum mopane,* occurring either as mopane woodland or scrub in dry, hot, low valleys up to about 1,000 meters a.s.l., and *Adansonia digitata,* the baobab, are also important and occasionally characteristic.

Evergreen communities or individual evergreen trees are fairly common in areas which are broadly referred to as savannas and savanna woodlands or woodlands. For example, among 84 common trees of Zambia, not including palms and euphorbias, 32 percent are deciduous, 49 percent semideciduous, and 19 percent evergreen. Evergreens occur largely in riverine environments, gallery woodlands, and a few other relatively well-watered deep-soil situations. The genera represented include *Parinari, Trichillia, Diospyros, Syzygium, Baphia, Cryptosepalum,* and others.

Deciduous and semideciduous thickets are common. Among the many genera of small trees and shrubs found in thickets are *Combretum, Commiphora, Balanites, Pseudoprosopsis, Bussea,* and *Pseudolachnostylis.* In wetter areas, evergreen thickets occur.

The region is of course famous for the diversity of its vertebrate fauna: Over 115 major mammal species, not counting mice, squirrels, and bats, many hundreds of bird species, and a fair number of reptilian and amphibian species occur on the plateau. Bovids are represented by about 50 species. The buffalo is found in all types of habitat, with the exception of large waterless tracts.

It is not clear to what extent the distribution of vegetation on the high plateau has been affected by man-made fires. Burning may have gone on, by design or semiaccident, for as long as man has known fire, and this may have been an important modifying factor in the size and distribution of grasslands and therefore in the ecology of grazing animals.

Specific areas

Three study areas are particularly often mentioned in the chapters that follow: Lolgorien, Kenya; Busanga, Zambia; and Mabalauta, Zimbabwe (Rhodesia). All three furnished important data, and they represent three very different buffalo habitats. Later comments may be more meaningful if these areas are summarized at the start. There will follow a few very brief notes on the other areas of study.

Lolgorien. Lolgorien is situated in the extreme west of the Kenya Maasai lands, 25 kilometers due west of the Mara river and 22 kilometers due north of the Tanzania border. It is located on a belt of Precambrian metamorphic rocks sculptured by erosion to form a number of 135- to 165-meter-high hills strongly elongated in a northwesterly direction and often topped with forest. Strips of forest also grow along watercourses in the bottoms of valleys as well as in the vicinity of the not distant and perennial Migori river. The long hill slopes that comprise most of the area are grass-covered, with small patches of forest, often only a few tens of meters in diameter, scattered over the grassland.

Some 3 kilometers east of Lolgorien and extending to the Mara river is a relatively high lava flat – the Isuria Plateau – overgrown by a patchwork of forest and grassland. From this lava flat the hills around Lolgorien are seen from slightly above. About 2½ kilometers south of Lolgorien and trending parallel to the Tanzania border, that is, along a northwest line, commences a large granite area. The landscape over the granite near Lolgorien consists of low rolling hills and depressions without any definite trend, covered by grassland and shrubs with occasional woodland patches and strips of riverine forest. Bounding the Lolgorien area on the west and north sides, 6½ to 11 kilometers distant, is the curving Migori River.

At the time of the study (1964–67) the human element consisted only of a small police post, a store catering mainly to the pastoral Maasai, a small government infirmary, and a small primary school. Several pastoral Maasai compounds separated from one another by a kilometer or so were scattered over some of the area. A handful of other people, unconnected with the above, and usually fewer than 10, inhabited the location more or less intermittently.

The fauna in general was rich. The bird and reptilian faunas were large and biogeographically typical. The abundance of the larger mammalian fauna was very evident. Besides buffalo, resident bovids included topi (*Damaliscus korrigum*), kongoni (*Alcelaphus buselaphus cokei*), waterbuck (*Kobus ellipsiprimnus/defassa*),[1] impala (*Aepyceros melampus*), reedbuck (*Redunca redunca*), bushbuck (*Tragelaphus scriptus*), oribi (*Ourebia ourebi*), red and Grimm's duiker (*Cephalophus natalensis* and *Sylvicapra grimmia*), and roan antelope (*Hippotragus equinus*). Eland (*Taurotragus oryx*) often passed through. Conspicuously absent were Thomson's and Grant's gazelles (*Gazella thomsoni* and *G. granti*), wildebeest (*Connochaetes taurinus*), and giraffe (*Giraffa camelopardalis*), four very common species on the east bank of the Mara River.

The area may have been the westernmost bastion of the black rhino (*Diceros bicornis*) in Kenya. Seven rhinos were resident within easy walking distance and others occurred in the vicinity. Zebras (*Equus burchelli*) and warthogs (*Phacochoerus aethiopicus*) were common. Elephants (*Loxodonta africana*) were usually present, especially just to the east and northeast of Lolgorien, and were occasionally seen browsing next to Maasai cattle when the latter grazed near trees.

Hyena (*Crocuta crocuta*), leopard (*Panthera pardus*), lion (*P. leo*), and cheetah (*Acinonyx jubatus*) were either resident or common visitors. Wild dogs (*Lycaon pictus*) also visited the area, and I observed them raising pups there. Serval (*Felis serval*), common and large-spotted genet (*Genetta genetta* and *G. pardina*), civet (*Viverra civetta*), side-striped jackal (*Canis adustus*), baboon (*Papio anubis*), and a number of small mammal species were also seen.

The Lolgorien area merited the status of a game reserve, open to cattle grazing and a few other activities, along the lines of Ngorongoro or the Maasai–Mara reserves. It was, however, one of the gazetted shooting blocks.

Busanga. The Busanga flats are situated in northwestern Zambia some 65 kilometers south of Kasempa and are mostly within the Kafue National Park. Even outside the national park boundaries the region is very sparsely populated and the land is a game management area.

The Busanga area includes woodlands and *dambos* but consists very largely of the vast floodplain of the Lufupa and Lushimba rivers which drain into the Kafue, and thus eventually the Zambezi.

The reader is referred to Chapter 12 for details on this area, where it is compared to the habitat of the Asiatic buffalo (*Bubalus bubalis*) resettled in Australia.

The Busanga area supports a large bird, reptilian, and amphibian fauna and a highly diversified medium-density mammalian fauna.

---

[1] Individuals resembling either species occurred.

During my working visits and stays in the Busanga area, between 1970 and 1978, I positively indentified many mammal species. Within the confines of the floodplain, including a thicket and several isolated "palm mounds," I encountered in different seasons the following species, besides the prominent buffalo:

Puku (*Kobus vardoni*), red lechwe (*K. leche*), roan antelope (*Hippotragus equinus*), situtunga (*Tragelaphus spekei*), reedbuck (*Redunca arundinum*), wildebeest (*Connochaetes taurinus*), eland (*Taurotragus oryx*), zebra (*Equus burchelli*), hippopotamus (*Hippopotamus amphibius*), warthog (*Phacochoerus aethiopicus*), bushpig (*Potamochoerus porcus*), rhinoceros (*Diceros bicornis*), elephant (*Loxodonta africana*), side-striped jackal (*Canis adustus*), wild dog (*Lycaon pictus*), hyena (*Crocuta crocuta*), lion (*Panthera leo*), leopard (*P. pardus*), cheetah (*Acinonyx jubatus*), serval (*Felis serval*), caracal (*F. caracal*), honey badger (*Mellivora capensis*), spotted-necked otter (*Lutra maculicollis*), clawless otter (*Aonyx capensis*), civet (*Viverra civetta*), genet (*Genetta genetta*), marsh mongoose (*Atilax paludinosus*), banded mongoose (*Mungos mungo*), other types of mongoose, cane rat (*Thryonomys swinderianus*), and vervet monkey (*Cercopithecus aethiops*).

On the periphery but overlapping with the floodplain I identified (besides many of the above species): oribi (*Ourebia ourebi*), Grimm's duiker (*Sylvicapra grimmia*), yellow-backed duiker (*Cephalophus sylvicultor*), waterbuck (*Kobus defassa*), sable antelope (*Hippotragus niger*), hartebeest (*Alcelaphus lichtensteini*), and Cape hare (*Lepus capensis*).

In woodland adjacent to the floodplain (again besides many of the species named above) I identified: kudu (*Tragelaphus strepsiceros*), bushbuck (*T. scriptus*), impala (*Aepyceros melampus*), baboon (*Papio cynocephalus*), wild cat (*Felis lybica*), antbear (*Orycteropus afer*), blue duiker (*Cephalophus monticola*), grysbok (*Raphicerus sharpei*), squirrel (*Heliosciurus* spp.), and porcupine (*Hystrix africae-australis*).

The Busanga flats and vicinity are one of the important remaining wildlife areas of Central Africa.

Mabalauta The Mabalauta area is situated in the far southwest of Zimbabwe, near the intersection of the Mozambique border with the Nuanetsi River. It is not on but peripheral to the African high plateau, in the so-called *lowveld*.

At the time of the study (1972) the area was a part of the Gona-Re-Zhou Game Reserve and uninhabitated except for the sector headquarters compound. Visitors were rare.

The area is largely underlain by crystalline rocks. During the dry season the streams dry out, with the possible exception of a few pools in the Nuanetsi riverbed and a few *pans* in relatively impervious depressions. The country is flat to hilly scrubland with some deciduous woodland patches and dotted with a few baobabs. Evergreen woodland is found along the Nuanetsi. The grass cover is generally medium high, with much bare ground showing between tufts. Some short grass occurs along the Nuanetsi and elsewhere.

The area does not support a very large mammalian fauna, as compared, for example, to Lolgorien or Busanga. Besides the buffalo, the common bovids were impala (*Aepyceros melampus*), kudu (*Tragelaphus strepsiceros*), nyala (*T. angasi*), bushbuck (*T. scriptus*), grysbok (*Raphicerus sharpei*), and waterbuck (*Kobus ellipsiprimnus*). Warthogs (*Phacochoerus aethiopicus*) and baboons (*Papio anubis*) were seen, and elephants (*Loxodonta africana*) were common. Lions were the most commonly seen predators. Leopards and black-backed and side-striped jackals (*Canis mesomelas* and *C. adustus*) were also present. There was also a variety of smaller mammal species, including banded mongoose (*Mungos mungo*).

Other areas. Other areas that furnished data at various times in the 1960s and 1970s are listed below and briefly annotated.

> Kenya coastal hinterland (southern areas)
> Samburu district, Kenya
> Tsavo National Park, Kenya
> Laikipia, Kenya
> Ruaha National Park, Tanzania

Dry: grassland – scrubland – thicket – woodland. Severe dry season during which water becomes highly localized. Good grazing and widespread water in the rainy season and for some time thereafter. Topographically variable but large flat areas are present in all locations.

> Luangwa valley, Zambia
> Zambezi valley, Zambia and Zimbabwe

Dry: open woodland (largely mopane) – sparse to fairly dense grassland – thicket. Severe dry season during which water becomes very localized. Good browsing and grazing and widespread water in the rainy season and for a time afterward. Broad-valley topography.

> Serengeti region, Tanzania
> Loita region, Kenya

Diversified: open grassland – woodland – thicket – evergreen forest. Relatively well watered on the whole, although some portions become waterless during the dry season. Flat to hilly, with areas of true hill country.

> Mount Kenya, Kenya
> Aberdare range, Kenya
> Ngurdoto crater, Tanzania

Montane: gladed forest – grassland patches – swamp. Access to pooled or seepage water.

> West Lunga, Zambia
> Lunga-Luswishi, Zambia
> Lukwakwa, Zambia
> Mkushi area, Zambia

Deciduous woodland – savanna woodland – *dambo*. Water variably localized during dry season depending on its severity. Flat to hilly topography.

2

Evolutionary background

In regard to the animal commonly known as the "Cape" Buffalo, naturalists, sportsmen, and scientists have evinced a ruling passion for creating numerous geographical races on the most slender evidence. (*Uganda Game Department annual report for 1925.*)

Il semble, lorsqu'on examine les variations existant chez les autres especes animales de l'Afrique, qu'il s'agisse d'Ongulés, de Rongeurs, de Carnassiers, qu'aucun autre mammifere ne presente une telle variabilité de forme que le Buffle. (*Paul Dalimier, 1955.*)

The emergence of the bovids

Ruminant ungulates probably originated in mid-Tertiary times (Late Oligocene or Early Miocene). The tribe Bovini, of the family Bovidae and subfamily Bovinae, which includes the African buffaloes, was the last division to evolve. The early bovids were small, antelopelike ruminants that probably lived in grassland and bush as do some modern duikers (e.g., *Sylvicapra*). They had already acquired high-crowned teeth, were able to ruminate, and the skeleton was essentially that of a modern advanced artiodactyl. Bovids had become one of the more successful mammalian herbivore groups during the Miocene [24.0 to 7.0 m.y.a. (million years ago)] and were numerous and varied. At the close of the Tertiary period (late Pliocene deposits dated at about 3.0 to 2.0 m.y.a.) fossil remains of bovids become increasingly numerous, and the first representatives of present-day genera are encountered. Subsequently, bovids increased still further in numbers and variety to become the dominant ungulate group in contemporary Eurasia and Africa.

13

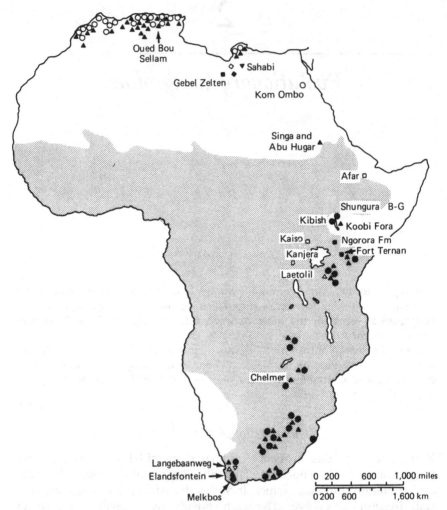

Figure 2.1. Sites from which fossil buffaloes and related bovids have been recovered in Africa. Shaded area indicates probable range of *Syncerus caffer* before extermination by man in some regions. Explanation of symbols: O, *Bos primigenius* (aurochs) or *Bos taurus* (cattle); ▲, *Pelorovis oldowayensis* and *P. antiquus* (long-horned buffaloes); ●, fossil *Syncerus* spp. (African buffaloes); △, *Simatherium kohllarseni* or sp.; □, *Ugandax gautieri*; ■, *Protragoceros* spp.; ◆, *Eotragus* sp.; ▽, *Mesembiportax acrae*; ◇, *Miotragocerus* sp.; ▼, *Leptobos syrticus*. The distribution of the known localities shows that *Pelorovis* was pan-African, *Bos primigenius* confined to the Mediterranean region, and *Syncerus caffer* sub-Saharan in Pleistocene times and chiefly known from the savannas. Plots are mainly from Hopwood and Hollyfield (1954) and Cross and Maglio (1975).

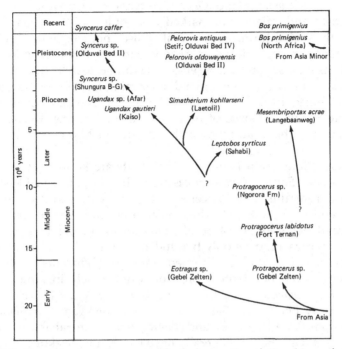

Figure 2.2. Possible evolutionary and phylogenetic relationships of African Bovini and related taxa (after Gentry, 1978).

The origin of the African buffalo, Syncerus caffer, and the fossil buffaloes of Africa[1]

The ancestry of the large cattlelike bovids lies in the Miocene artiodactyls that lived probably in Asia. The African record is sparse and often comprises isolated teeth or postcranial elements that can only be identified to the Bovidae. What evidence there is comprises a few skulls, one or two incomplete skeletons, and some horncores and limb bones from sites in South Africa, the Rift Valley areas of East Africa, and the Mediterranean coast of the Maghreb (Figure 2.1). This account is based chiefly on that of Gentry (1978), whose understanding of the evolution of the Bovidae is second to none.

The first African bovids occur in the Miocene deposits that antedate those at Fort Ternan, Kenya, from which the first ancestral cattlelike forms are known (Figure 2.2). Hamilton (1973, pp. 126–8) recorded *Protragocerus*, *Eotragus*, and *Gazella* from Gebel Zelten, Libya (Early Miocene, 17 m.y.a.); *Eotragus* may lie close to the ancestry of the later bovids. Fort Ternan (Middle Miocene, 14 m.y.a.) has yielded bovids

[1] This section was contributed by C. S. Churcher.

(*Protragocerus labidotus* and *Oioceros tanyceras*) with wide low skulls, brain cases and palates not markedly out of line (little cranial flexion), frequently keeled horncores, frontal sinuses not frequent, cheek teeth with rugose enamel and not markedly high-crowned (hypsodont), persistent basal pillars on molars (endostyles and ectostylids), long premolar rows, shallow horizontal mandibular rami, and not highly cursorial limbs (Gentry, 1970, pp. 245–6, 277–82). The boodont bovids known from the Siwalik deposits of India have a similar series of characteristics, and may have affinities with the African Miocene and Pliocene bovids.

In Africa four main lineages of boodonts are known. *Leptobos* from the Miocene, *Simatherium-Pelorovis* from the Pliocene and Pleistocene, *Syncerus* from the Late Pliocene to the present, and *Bos* from Late Pleistocene to the present. These genera are members of the Bovini, large-sized descendants of boselaphine bovids of the Tertiary, which are now represented by only two Indian species, the nilgai (*Boselaphus tragocamelus*) and the four-horned antelope (*Tetracerus quadricornis*). Members of the Boselaphini once occurred widely in Africa, for example, *Mesembiportax acrae* from Langebaanweg, Cape Province, South Africa (Early Pliocene to Early Pleistocene), *Protragocerus labidotus* from Fort Ternan (Late Miocene), and *Miotragocerus* from Sahabi, Libya (Late Miocene). Living Bovini generally have low and wide skulls, horncores in both sexes that emerge transversely, internal sinuses in the frontals and extending into the horncores, short braincases and long facial regions, triangular basioccipitals, molars with basal pillars and complex enamel lakes (fossettes), upper molars with strong convex buccal ribs on the paracones and metacones between the styles, and no parastylid fold (goat fold) on the lower molars.

Syncerus is distinguished from the other Bovini by its short facial region, short horncores arising from just posterior to the orbits, and the fusion of the paraconid to the metaconid to close the medial wall of the lower fourth permanent premolar. It is highly variable in many of its characteristics, as are *Bos*, *Bison*, and *Bubalus*, and identification to genus from single teeth or fragmentary remains is often unreliable, even when all structures are represented but not as diagnostic variants (cf. Gentry, 1967, p. 277, Fig. 11, and p. 279).

Petrocchi (1956) described *Leptobos syrticus* (Figure 2.3a) from Sahabi, Libya (Late Miocene, perhaps 6 m.y.a.) on three crania and other elements. The cranial characters suggest that *L. syrticus* is primitive when compared to Asian or European *Leptobos* in that the horncores lie closely posterior to the orbits, they have short bosses or pedicels, anterior and posterolateral keels are present, and the suprorbital pits are placed close to the sagittal line. In these characters it resembles *Parabos boodon* from France, but probably represents a separate lineage (Gentry,

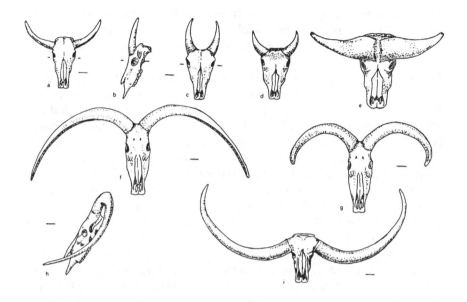

Figure 2.3. Horn conformations in African Bovini. (a) *Leptobos syrticus*, recon-structed from Petrocchi (1956) – areas anterior to the orbits are hypothetical; (b, c) *Ugandax gautieri*, reconstructed from Cooke and Coryndon (1970, plates 17C, 18D) – Areas anterior to the orbits are hypothetical; (d) *Syncerus caffer nanus*, based on Lydekker (1926, plates III.2, III.3, III.5); (e) *S. c. caffer*, male, based on Lydekker (1926, plate III.1) and Roberts (1951, plate 29.1); (f) *Pelorovis oldoway-ensis*, skull from Olduvai Bed II in the National Museum of Tanzania, originally assigned to *Bularchus arok* by Gentry (1967) but reassigned by Gentry (1978), redrawn from Gentry (1967, plate 5, fig. 3); (g) *P. oldowayensis*, skull Pel 1, from Olduvai Bed II, redrawn from Gentry (1967, Plate 1, fig. 2); (h) *P. oldowayensis*, reconstructed lateral aspect; (i) *P. antiquus*, redrawn from Gentry (1978, p. 565, fig. 27.9B). Horizontal bars indicate 10 cm. Specimens a to d drawn to same skull length and e to i to greater skull length to show similarities between *U. gautieri* and *S. c. nanus*, and between *P. antiquus* and *S. c. caffer*.

1978 p. 549). *L. syrticus* is also unlikely to be related directly to *Syncerus* because the area between the horncore bosses is smooth rather than rough, the roof of the braincase is not sufficiently sloping, the temporal ridges are overly developed, and the posteromedial keel on the horn-cores is weak. However, *L. syrticus* is the earliest bovid that is a possi-ble ancestor for *Syncerus*, and for it to be on the direct lineage some reversion of evolutionary trends would have to have occurred or else the Sahabi specimens would have to represent aberrant samples.

The earliest recognized member of the *Syncerus* lineage appears to be *Ugandax gautieri* (Figure 2.3b, c) from Kaiso, Uganda (Cooke and Co-ryndon, 1970 p. 206) of unknown age. This was first thought to be a hippotragine antelope but Gentry (1978) reassigned it to the Bovini.

The braincase is higher and longer than in later members of the lineage, and the horncores are smaller and less dorsoventrally flattened and are oriented to pass dorsally, then posteriorly, and finally medially, quite different from the orientations of the horns of *Syncerus*. Gentry (1978, p. 548) considered *Ugandax* to resemble *Proamphibos* from the Siwalik Tatrot Formation (Pliocene), which is thought to be ancestral to the Asiatic *Hemibos* sp. and modern water buffaloes (*Bubalus* sp.), but it differs in its shorter horncores with weaker keels, a more parallel-sided basioccipital, and smaller basal pillars, styles, and ribs on the upper molars. A more advanced *Ugandax* is known from Afar, Adar Formation, Ethiopia (Late Pliocene).

Syncerus is known from the Shungura Formation, Members B–G, Ethiopia (Pliocene–Early Pleistocene). This *Syncerus* has a lower and shorter braincase than *Ugandax*, strong temporal ridges, and short horncores. In Olduvai Gorge, Tanzania, in the middle and upper parts of Bed II (Late Early or Early Middle Pleistocene) a *Syncerus* species occurs that is probably ancestral to the modern *S. caffer* (Figures 2.2 and 2.3d, e). Its crania have horncores that are hollowed only near the base and are triangular in section with anterior, dorsal, and ventral surfaces; the fourth lower permanent premolars have closely opposed but unfused paraconids and metaconids and a wedge-shaped basioccipital that is less tapered than in *S. caffer*. The horns are oriented transversely and lack large basal bosses, as in early *S. caffer*.

Modern *S. caffer* (Figure 2.3e) south of Ethiopia has convexly arched frontals, large basal bosses to the horncores, and down-turned horncores in adulthood. None of the more recent occurrences of fossil *Syncerus* show individuals with these features and thus they may have originated relatively late in the evolution of the genus. An almost entire skeleton assigned to *S. caffer* has been recovered from the Kibish Formation, Omo, Ethiopia (Leakey, 1969, p. 1132; Late Pleistocene–Recent) and resembles the less-advanced populations of *S. c. aequinoctialis* of the West African and Ethiopian savannas, but is as large as a large bull of the southern *S. c. caffer* type. In South Africa, horncores of an individual from Melkbos, Cape Providence (Hendey, 1968, p. 104; Late Pleistocene, 200,000–130,000 B.P.) have the hollowed horncores and rugose frontals, as in modern *S. caffer*, but the horns pass laterally and do not run immediately ventrally. The impression is thus of much possible variation (Grubb, 1972), both individually within a population and between populations at any one time in the Late Pleistocene, such that both modern and ancient morphotypes might have been present in the extant herd. Such a condition appears to have existed within the North American bison (*Bison* spp.) as it evolved from the long-horned "*B. priscus–B. latrifrons*" condition to the modern "*B. bison*" form (Wilson, 1975).

The known distribution of *Syncerus* is restricted to sub-Saharan Africa, and no extant or extinct records of *Syncerus* have been verified from North Africa, the central Saharan massif, or the Nile Valley north of Nubia. This, together with the observed geographic variation and probable evolutionary history, suggests that *Syncerus* is an African endemic that originated in the area between Lake Victoria and the Horn of Africa, either on the Great Plains of East Africa or in the Ethiopian Highlands. From there it spread in two directions, westward and southward, to give rise to the modern *S. caffer aequinoctialis* in the west and *S. c. caffer* to the south, respectively. It appears that *S. c. aequinoctialis* represents a more conservative lineage and *S. c. caffer* a recently derived and more differentiated conformation.

There also exists evidence for a lineage of large-horned buffaloes in the Middle and Late Pleistocene of northern, eastern, and southern Africa. The first recognizable material was named *Simatherium kohllarseni* by Dietrich (1941, p. 22; 1942, p. 119) who obtained a badly preserved cranium from Laetolil, Tanzania (Pliocene). Langebaanweg, Cape Province, South Africa (Early Pliocene–?Early Pleistocene) has yielded similar fragments (Gentry, in Hendey, 1970, p. 114), but the horncores have very prominent keels similar to those in the Shungura *Syncerus* material (Gentry, 1978 p. 549) or the irregular keels present in the Elandsfontein *Pelorovis* (see below). *S. kohllarseni* may be ancestral to *Pelorovis*, the best-known genus of the long-horned African buffaloes, but differs from it in that the horncores arise from closer to the orbits, are more widely spread, and diverge less toward the tips.

The later members of the large-horned lineage may be conveniently grouped within a single genus, *Pelorovis*, and may constitute a natural group. Two species are recognized, the Early Pleistocene *P. oldowayensis* (Figure 2.3f–h) and the Late Pleistocene *P. antiquus* (Figure 2.3i), and possibly a southern variety from Elandsfontein, Cape Province (Middle Pleistocene), which is yet insufficiently documented to warrant either inclusion within a known species or separate status. These buffaloes differ from *Syncerus* in their large, arcuate horncores, which are less dorsoventrally compressed and lack basal bosses, and in their longer metapodials.

Bate (1949, 1951) described a new genus and species, *Homoioceras singae*, from Singa and Abu Hugar, Sudan (Late Pleistocene), based on a skull with long horncores and bases. Bate intended her genus to include all the long-horned African buffaloes and to distinguish them from the Asiatic water buffaloes of the genus *Bubalus*. Gentry (1978, following a suggestion by J. W. Simons) recognized *H. singae* as a large individual of the *Syncerus* lineage because of its markedly dorsoventrally compressed horncores and their origins close to the midline on the skull roof. *H. singae* is thus *Syncerus singae* or a synonym of *S. caffer*. *Pelorovis* (Reck, 1928) is a possibility for the large-horned lineage, which

as Bate correctly recognized, is distinct from Asiatic *Bubalus* and resembles *Syncerus* in shorter facial region, has reduced, absent, or ir-
regularly developed keels on the horncores, short nasal-premaxillary
sutures, nasals unflared anteriorly, a low skull with wide occiput,
wedge-shaped basioccipital, no fusion between the vomer and the
palatine, and a tendency for the paraconid and metaconid on the lower
fourth permanent premolar to fuse.

The deposits in Olduvai Gorge have yielded large-horned buffaloes,
Pelorovis oldowayensis Reck 1928 (including *Bularchus arok*: Hopwood,
1936 – see Figure 2.3f; Gentry, 1967) from the middle and upper levels
of Bed II, although the holotype may derive from Bed IV. *P. oldowayensis* is also known from Kanjera (Early–Middle Pleistocene) and high in
the Koobi Fora Formation, East Turkana, Kenya (Middle Pleistocene;
Harris, 1976, p. 295). The holotype has horncores set close together
and posteriorly above or near the occiput and which curve posteriorly,
then laterally, and finally ventrally from the skull. The teeth are distinct from those of *Syncercus* from Olduvai Bed II in their larger size
and simpler enamel pattern (Gentry, 1967, p. 277, Fig. 11). Such distinctions are less obvious when the Shungura *Syncerus* is compared
with the Olduvai *Pelorovis*, as might be expected in populations possibly closer to a common ancestry.

The Late Pleistocene species is *P. antiquus*, first described from Oued
Bou Sellam, Algeria, by Duvernoy (1851; Figure 2.3i) and in greater
detail by Pomel (1893) as *Bubalus antiquus*. Pomel also described other
North African specimens now assigned to this species. Other records
exist under the names *Bubalus baini*, from Chelmer, Zimbabwe and elsewhere in South Africa (Cooke, 1949) and *B. nilssoni* from the Malewa
River, near Naivasha, Kenya (Lonneberg, 1933; Nilsson, 1964) and appear to be chiefly confined to the South and East African savannas. *P.
antiquus* differs from earlier *P. oldowayensis* in having more complex enamel patterns on the cheek teeth and less ridged horncores, except for the
specimen recovered from Elandsfontein, Cape Province, which may
represent a population divergent from *P. oldowayensis*. The horncores of
P. oldowayensis curve backward and upward on leaving the skull, and
those of *P. antiquus* usually curve downward and laterally. The origins of
the horns moved anteriorly to just behind the orbits as in *Simatherium*
and probably reflected a readjustment of the metacentric balance of the
skull, with the center of gravity being lowered to below the condylar
articulation and forward, rather than above and behind with a resultant
constant tendency for the noses to be raised and the instability of a high
center of gravity removed. Other changes that appear to have taken
place are the shortening of the facial region of skull anterior to the orbits,
more protuberant orbits, and a broadening of the nuchal plate, so that
the skull of *P. antiquus* is broadly similar to that of *Syncerus*.

The large-horned buffaloes became extinct by the beginning of the

Holocene and records of *"Homoioceras"* from Kom Ombo (*Bubalus vig-nardi*–Gaillard, 1934; Churcher, 1972, 1974), once considered as possible *Pelorovis*, are now considered to represent large aurochs, *Bos primigenius*, or, less likely, *Syncerus*.

The third large bovine present in the Quarternary of Africa is the aurochs, *Bos primigenius*. It is well known from Late Pleistocene or Early Holocene deposits in North Africa and has been replaced ecologically by domestic cattle, *Bos taurus*, which are derived from it. Aurochsen are known only from deposits north of the Sahara, and chiefly along the Mediterranean littoral (Churcher, 1974), the oases of the Western Desert (Churcher, 1980, 1981), or the Nile Valley (Churcher, 1972), and provide the ecological substitute for the Cape buffalo in similar latitudes. Pastoral herding of cattle in the savannas of Africa over the past few millennia, culminating in the accelerated and European-inspired introduction of *B. taurus* varieties to the sub-Saharan savannas, has reduced the range and numbers of African buffaloes over the last few hundred years.

The cause or causes of the extinction of the large-horned buffaloes is unknown, but was not competitive exclusion by *Bos*, since *Bos* coexisted with *Pelorovis* (=*Bubalus*) in North Africa (Churcher, 1974, p. 377), and *Pelorovis* died out in sub-Saharan Africa when *Bos* was absent. It may be speculated that other large herbivores provided competitive exclusion, for example, *Taurotragus* at Klassies River Shelter (R. G. Klein, personal communication, 1979), especially the expanding populations of *Syncerus*, or there may have been predatory overkill by Holocene man, or that an epizootic disease occurred to which they had insufficient resistance.

It may also be suggested that *Pelorovis* evolved as a savanna dweller in East Africa and spread throughout the African grasslands. At the same time *Syncerus* may have evolved parallel adaptations but in the Ethiopian highlands where woodland was more prevalent and forests existed. With the emergence of modern *S. caffer*, the Cape buffalo lineage acquired a form that was adapted to the savannas as well as forests and was able to outcompete *Pelorovis antiquus*. Thus *P. antiquus* might have been unable to coexist with the more efficient *S. caffer* in sub-Saharan Africa and succumbed to competition from *Bos* and *Homo* in combination in North Africa. However, as the situation is presently understood, all answers are speculative when reasons for the extinction of *Pelorovis* are contemplated.

How many species? The taxonomic history of Syncerus caffer

At least 92 zoological names have been given to African buffaloes since 1779, when the name *Bos caffer* Sparrman was established, and many of them were thought to represent separate species. These species were reduced to three by Sir Victor Brooke (1873, 1875), who provided the first rational framework for the classification of the African buffaloes, which until then had lacked an overall perspective of their interrelationships.

The degree to which buffaloes vary from individual to individual, even within one herd, was not fully appreciated until the present century. Horn shape, one of the most variable features in buffaloes, is also one of the most easily observable, and it was often the only reason for considering specimens as zoologically distinct. This preoccupation with the shape of the horns was intensified in the early days by the shortage of other available features for comparison, since the early specimens seem to have consisted only of skulls and/or skins. Furthermore, most of them came from West Africa, where variation among buffaloes is great. As a result, previous to this century the supposed number of buffalo varieties was not much smaller than the number of specimens available for study. Horn configurations have been the main characteristic used in some attempts to subdivide the African buffaloes as recently as the middle of the present century (e.g., Roberts, 1951). Horn shapes and sizes are considered less significant in recent efforts to classify the African buffaloes, although they still figure as one of the important criteria (e.g., Grubb, 1972). Dalimier (1955) remarks that the Indian buffalo (*Bubalus bubalis*) shows an extraordinary diversity of horn shapes and implies that unequal taxonomic importance is being attached to this feature in the two genera of living buffaloes, *Bubalus* and *Syncerus* – that is, that horn shape is less taken into account in the Asiatic genus.

By 1872 there were no fewer than 18 references to the African buffalo in the European scientific literature, although several of them were based on one and the same source material. It was becoming apparent that the buffaloes of southern and southeastern Africa were less physically diversified than the western and west-central African buffaloes. It was also gradually becoming apparent that many of the names coined by the eighteenth- and nineteenth-century naturalists for the more physically diversified western and west-central African buffaloes, based largely on horn configurations, were synonymous. Accepted as synonymous were *Bos nanus* Boddaert, *Bos pumilus* Kerr, *Bos corniculatus* Blyth, *Bos planiceros* Blyth, *Bubalus reclinis* Gray, *Bubalus brachyceros* Gray, and others varying in zoological respectability.

Brooke's two papers (1873, 1875) dealt largely with the work of his predecessors, but he deserves the credit for reducing a chaotic situation to one of order. First, he indicated that sharp breaks between African buffalo types should not be expected. Second, he assigned all the African buffaloes to only three species: *Bubalus pumilus*, *B. caffer* Sparrman, and *B. aequinoctialis*. The first species comprised all the small western and central African buffaloes; the second, all the large southern and southeastern buffaloes; the third, the medium-sized buffaloes of northeastern sub-Saharan Africa, which are intermediate in several respects but whose mode of life approaches that of the Cape buffalo. The "Conclusion" of Brooke's second paper (1875, p. 457) is quoted in full, as it is a taxonomic milestone:

In conclusion I would simply say that although for the moment it seems to me decidedly advisable to regard the three forms of African Buffaloes as distinct species, each known by a separate name, I am fully aware of the slender basis upon which their distinctive characteristics rest. I have, indeed, already seen specimens of Buffaloes from the Upper Zambesi of strikingly intermediate characters between *Bubalus caffer* and *Bubalus aequinoctialis*. Between this latter form and *Bubalus pumilus* the difference of external characters appears to signify a wider breach; but that these superficial differences may be found deceptive is, I think, rendered probable by the very remarkable gradation of characters taking place more or less step by step with the gradual spread of the animals over their geographical range.

Even afterward, superficial differences had a way of prompting investigators to proclaim new species and races. Some important differences may have really existed, however, although now it may be too late to raise them above the speculative level, owing to extinctions. Buckley (1877) wrote:

Mr. Du Bois, with whom I was hunting . . . in the Anaswazi country, and who knows that part of Africa perhaps better than any other man, informs me that a variety of the Buffalo smaller and with a red tinge on its skin, used to exist along the Bomba hills; and in fact I saw such a skin brought in for sale by one of the natives. The Hon. W. H. Drummond, in his book on the Large Game of Southeastern Africa, says, p. 33, 'A herd of Buffalo, or, more correctly speaking, several herds, that exist in a district known as the Umbeka, on the northeast of Zululand, are famed as having a tinge of red in their colour, and as being smaller and more dangerous than any other.'

It was, however, becoming apparent that very few, if any, types of buffalo found in southern and southeastern Africa diverged substantially from Brooke's *Bubalus caffer* Sparrman. Both *Bos* and *Bubalus* continued to be used as the generic name for the African buffalo. It appears that the generic name *Buffelus* (e.g., *Buffelus caffer* Roberts) was proposed in 1914, although it did not find acceptance.

Lydekker (1913), who had been studying bovids since the 1890s, expressed the view that all African buffaloes belonged to a single species. It was an escalation of Brooke's thinking. But the fact remained that African buffaloes came in a large variety of sizes, shapes, and hues. Students of buffaloes remained uneasy about placing them all in one species. A compromise between Brooke's three and Lydekker's one species was opted for by some zoologists, who settled for two species. Christy was a prominent supporter of this approach. In his 17-page paper (Christy, 1929), he implies two species, and ends (p. 459) with the following synopsis:

Group I. *Bubalus caffer* Sparr., Black (Cape) buffaloes.
Group II. *Bubalus nanus* Bodd., Red (Congo) buffaloes.
    Subgroup i. *Bubalus nanus nanus* (Intra-forest races).
      (a) *B. nanus nanus* var. *reclinis* Blyth
        (Lyre-shaped).
      (b) *B. nanus nanus* var. *planiceros* Blyth
        (Lobster-claw-shaped).
    Subgroup ii. *Bubalus nanus aequinoctialis* (Extra-forest races).

Christy states (1929, p. 457):

At one time all the forms of African buffaloes may no doubt have been races of a single variable species, but nowhere to-day in the extensive ranges of either *B. caffer* or *B. nanus* can heads of one species be mistaken for those of the other, except on the Upper Nile, where *aequinoctialis* and *B. caffer* meet, and where the gap between the horn characters of the two groups is narrowest. When size and coloration are taken into account, however, I believe the gap between the two groups is even here distinct.

The generic name *Syncerus*, which excludes all living buffaloes save the African types, was introduced by Hodgson (1847) and has gradually displaced the other generic names among the English-language authors; "*Bubalus*" is still encountered, however, in the literature of the last few decades, notably among French-language authors. The uneasiness about the classification of African buffaloes has not disappeared, although the majority of recent authors accepts a single species, *Syncerus caffer*, composed of two subspecies, *S. c. caffer* (the "Cape" buffalo) and *S. c. nanus* (the dwarf or forest buffalo).

Variable physical characteristics of Syncerus caffer

If a single-species classification is accepted, one must automatically accept a great amount of intraspecific variation; if more than one species is assumed, interspecific boundaries must be defined, despite gradational characteristics. The extreme types of the African buffalo are the great buffalo of the southern and southeastern regions, *S. c. caffer*, reaching a shoulder height of more than 1.6 meters and a weight of over 835 kilograms, and the little buffalo of the Ituri and Aruwimi forests of the Zaire (Congo) basin, *S. c. nanus*, standing sometimes only just over a meter at the shoulder and weighing under 300 kilograms. A gradation of types links these extremes. The occurrence of very wide variations within a single locality is a feature that makes diversification among African buffaloes difficult to deal with on a systematic basis. This condition occurs throughout the African buffalo's range but becomes most pronounced in its western portions. There are references to single herds in the Central African Republic (Edmond-Blanc, 1935; Reboussin, 1953), Togo (Baudenon, 1952), Gabon (Maclatchy, 1932), Zaire (Lydekker, 1908), and Nigeria (Henshaw and Greeling, 1973) that show great variations in the easily observable characteristics. A number of these variations, particularly in body color and horns, are attributable to differences in age and sex, but a substantial remainder cannot be so explained. My own observations, in Ghana as well as in eastern and southeastern Africa, support the existence of variation that is not due to age or sex differences. On the eastern side of Africa, among buffaloes of the *S. c. caffer* type, I noted an increase in local diversification as I traveled south-southwest, from northern Kenya to about 14°S latitude. From there on southward, there appears to be a decrease in diversification within local populations. However, individual diversification among African buffaloes is almost everywhere of noticeable proportions. Localities on the eastern edges of the Zaire (Congo) watershed, such as western Uganda or Zambia's Mporokoso district, tend to have markedly diversified buffalo populations, with diversification observable either within or between herds, or both.

The conclusion that all African buffaloes should be considered conspecific is very largely based on this geographic overlapping of diverse characteristics – size, color, horn shape, massiveness of body, and hair – despite the very large difference between the two extreme types. Dorst and Dandelot (1970, p. 274) state on this score: "For practical reasons it is best to consider them [the African buffaloes] as monospecific, though they probably derive from two different types which integrate on a large scale." The view that the present-day African buffaloes were derived from the merger of two quite distinct species, presumably still represented by the extreme types, was also expressed by Malbrant before the French Societé d'Acclimatation in 1935, but does violence to accepted concepts of the biospecies.

In the field of skeletal or cranial comparative anatomy, no one has found any conclusive evidence for deciding that all living African buffaloes comprise more than one species. For distinctions recourse must be had to tendencies, rather than sharp differences, such as the variation in the position of the suture between the maxilla and the premaxilla, with respect to the nasal bone (e.g., Dalimier, 1955) which, in my experience, is subject to exceptions: In *S. c. caffer*, the junction between the maxilla and premaxilla is not always in contact with the nasal bone, but may be anterior to the latter, as in *S. c. nanus*.

Despite the massive disappearance of local buffalo populations,

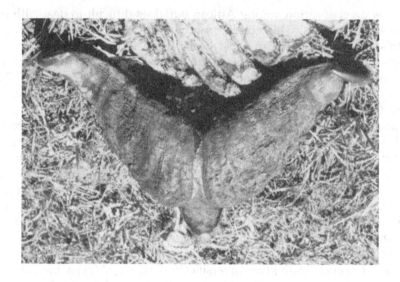

Figure 2.4. Comparison of mature horn development in the male (top) and female (bottom) of *S. c. caffer* var. *capensis*. (Western Zambia.)

through disease such as the rinderpest epizootics of the 1890s and the 1920s, extermination by hunters, or spread of settlement, gradations between the surviving types are usually self-evident. The gradations tend to parallel the density of tree cover that is typical for the habitats of various buffalo populations.

The smallest African buffaloes are the dense-forest type. Buffaloes of regions that are transitional between forest and bush/grassland are of intermediate size. Bush/grassland buffaloes are the largest. In forest buffaloes, a tendency appears to exist toward an increase in body size as the forest habitat departs from ideal deep forest. This can be observed if one compares, for instance, buffaloes of the Ituri forest in eastern Zaire with those from Equatorial Guinea.

If one takes the horns of a typical adult bush/grassland buffalo (*S. c. caffer* var. *capensis/radcliffei*) as one's standard for comparison, then the forest buffaloes have small light horns, relative to body size, with shapes not unlike juvenile phases of the standard form (Figures 2.4 and 2.5). Two early investigators, Pennant (1781) and Gray (1852), thought that certain horns, now accepted as belonging to adult West African buffaloes, were horns of young *S. c. caffer*. Transition-zone buffaloes have medium-sized horns. Similarities in shape exist between horns of some adult West African types – those colloquially known as bushcows – and horns of the young subadult Cape buffalo, as well as between adult Chad or Sudan buffaloes and subadults of the Cape race. Are we witnessing ontogeny in the various phases of horn development in a maturing Cape buffalo – or possibly pedomorphism in the horns of the dwarf buffalo? Grubb (1972, p. 124) showed, on the basis of 280 specimens (mainly male), that horn size tends to increase outward from the west-central African subequatorial forest zone, toward the north, east, and south. All types of *Syncerus* horns alter somewhat and coarsen with age, and the development of a basal horny thickening, very considerable and known as a boss in Cape buffalo males but only slight in forest types, is a general feature of old males. This horny boss is supported by a bony thickening, confusingly also referred to as the boss.

Forest buffaloes occur in groups of up to eight individuals, although deep-forest buffaloes of the Ituri type (*S. c. nanus* var. *nanus*) tend to smaller groups. Transition-zone buffaloes commonly occur in groups of up to 20, rarely 50, but in a few areas where true bush/grassland conditions were approached in the remembered past these groups increased to over 100 (verbal communications from two old native African hunters, 1962, 1963). The bush/grassland buffaloes occur in groups attaining many hundreds where primitive conditions obtain, occasionally exceeding 2,000.

The hair of the forest buffalo is light to dark reddish brown for life, or it may darken to almost black in advanced maturity. Transition-zone buffaloes are normally some shade of brown or dull brownish yellow in early adulthood, often darkening in advanced maturity to deeper, sometimes sooty brown, with or without a red tinge, to nearly black. Some are black throughout life. In Chad and some wooded parts of Uganda, fully adult buffaloes of a light reddish brown hue are, or were

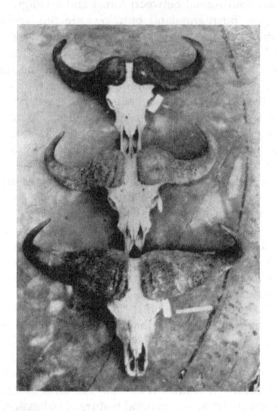

Figure 2.5. Horn development in three adult males of *S. c. caffer* var. *capensis*, all from one herd. (Busanga area, Zambia.)

Figure 2.6. This 8¼-year-old Chad male (*S. c. nanus* var. *brachyceros*) is covered with light reddish yellow hair except for nearly black "socks." The ears are heavily tufted. The horns lack any downward development. The animal's size is comparable to that of a female *S. c. caffer* var. *capensis/radcliffei*. (Photographed in the Antwerpen Zoo, 1976.)

until recently, common. Figure 2.6 shows a male Chad buffalo, photographed at the age of 8¼ years, whose entire body is covered with light reddish yellow hair. Littlejohn (1938, p. 17) states, referring to bush-cows from Akparabong on the Cross River in southeastern Nigeria: "Reliable native hunters said that the usual colour is red, but that black animals of either sex are occasionally seen. They stated that a black animal is black from birth, and one man said he had taken an unborn calf which was quite black from a cow he had killed." This statement, though interesting and possibly illustrative of the kind of individual diversity that is so often noted in buffaloes, requires comment: I have not seen a bushcow at birth, but I have observed that calves of the *S. c. caffer* type, which are often quite light brown when several weeks old may be very nearly black at birth. Perhaps that was the case with the unborn calf of Littlejohn's statement? Bush/grassland buffaloes of the Cape type have dark brown to nearly black hair, but in some regions, as for example from about Iringa in Tanzania to the vicinity of the Limpopo river, calves are commonly of varying shades of brown, often reddish and quite light. I have observed that, without notable exception, the lightest-colored calves are most thickly covered with hair that is relatively fine, occasionally giving a first impression of being woolly (Figure 2.7a). In bush/grassland buffaloes that were lighter-colored as

a

b

Figure 2.7. (a) A calf from the Busanga area, Zambia. The lightest-colored calves are in general most thickly covered with hair, occasionally giving a first impression of wooliness. The lighter hair persists on parts of the body until adulthood. The calf is grazing. (b) This calf is from the same population as the calf in (a). Its hair is sparser and nearly black. Calves that are born black may subsequently become lighter.

calves, the lighter hair persists on parts of their bodies until they reach adulthood. The same Cape buffalo populations that produce light brown calves also produce black ones: The calf in Figure 2.7b is from the same population as that in Figure 2.7a, that is, from the Busanga region of northwestern Zambia. There is no simple correlation with the mother's hair color. The skin color of the forest buffalo is generally gray, occasionally rather pale. The skin of the transition zone and bush/grassland buffaloes is gray to nearly black. Pinkish or "dirty white" blotches occur, varying in size from a centimeter or so to a meter or more across, and they may occur singly or in groups. Light skin blotches are not uncommon in transition-zone buffaloes but rare in bush/grassland types. A noteworthy case in the bush/grassland type was reported by Bainbridge (1969) from the Luangwa Valley, Zambia, of a subadult with a very pale ventral area, the pale color extending dorsally across the middle of the body in sharp contrast to the remainder of the animal. Those transition-zone buffaloes that retain rufous pelage into advanced maturity appear especially likely to have unevenly pigmented skin. Pale blotches on the nose have been observed in bushcows, Sudan buffaloes of the *aequinoctialis* variety, and Cape buffaloes. Although exceptions exist, there is a general trend from lighter to darker, starting with forest buffaloes, passing through transition-zone buffaloes, and ending with bush/grassland buffaloes.

A rough parallelism exists between the body size of buffaloes and the abundance of grasses in a given habitat. Grass is scarce and unevenly distributed in dense forest regions, largely restricted to places that can be reached by direct sunlight; in the transition zone as a whole grass is moderately plentiful but also somewhat unevenly distributed and there may be considerable competition for it; in bush/grassland grass is very abundant. It appears that the more grass there is in a region, the larger are the buffaloes inhabiting it. This is hardly surprising since all Bovini are equipped to be predominantly grass eaters.

There are then several features common to all buffaloes that are gradational from forest to transition-zone to bush/grassland types. The greatest diversification occurs in the transition-zone types, but that is due to the arbitrariness of this classification: Deep forest and bush/grassland buffaloes are end- or near-end-members of the series, whereas the "transition zone" is a repository for all the diverse intermediate types.

The presence of gradational characteristics and the lack of any fundamental differences among African buffaloes as a whole support Lydekker's view that they are conspecific. Furthermore, if a diversified species of grass eaters par excellence such as *Syncerus caffer*, where one end-member lives on grass, while the opposite end-member lives in and is adapted to a grass-poor habitat, the first hypothesis must surely be that the former end-member is probably closer to the ancestral type. It is conceivable that the little deep-forest buffalo is a stunted form owing to malnutrition.

The advocates of the view that the small buffalo is the ancestral form of *Syncerus* make much of the apparent fact that the central African forests used to cover a larger area in the past than they do at present (Dalimier, 1955). There is little doubt of this (e.g., van Zinderen Bakker, 1962). However, the shrinking of the subtropical forest in the time following the last glacial stage is only a part of the picture, and surely more important, from the viewpoint of living forms, is a related process, the desiccation of the northern African savanna which changed into desert. This savanna, still in existence some 3,000 years ago (van Zinderen Bakker, 1958, 1962) was probably in large part semiarid and unlikely to carry major populations of highly water-dependent species, such as *Syncerus*, although there is no climatological reason why they should not have been present along watercourses; and the region apparently did carry populations of less-water-dependent herbivores and browsers, such as some antelopes, giraffes, and even elephants, for which there is paleontological and archeological evidence (e.g., McBurney, 1960; Churcher, 1981). Even if *Syncerus* were always absent from the northern and central reaches of the Saharan region, it probably existed in the south of it, prior to desiccation, on grasslands which must have occurred in place of the approximately 475-kilometer-wide dry savanna belt of today, which intervenes between the Sahara and the central African forest region. The desiccation process, geologically extremely rapid but gradual in terms of mammalian generations, is unlikely to have caused catastrophic extinctions; it is much more likely to have caused large segments of the Saharan fauna to migrate southward, to where the remaining grasslands met the subequatorial forest. It appears probable that competition for these peripheral grasslands was great, as the indigenous fauna was augmented by migrants from the progressively more arid north. The more adaptable forms may easily have been faced with the choice of either accepting critically acute competition for food, and eventually water, in the congested grasslands, or migrating into the forest. Some of the surviving northern buffalo population may have taken the latter course. Under the new forest conditions, these buffaloes would be comparatively restricted, both in terms of food selection and of the genetic pool from which they had to breed, as they were now compelled to exist in small groups. Under such conditions, primitive and possibly rare or recessive characteristics, inadaptive in larger breeding populations, could be expected to reappear (e.g., Wright, 1922), including loss in fertility, which would tend to further reduce forest populations. Between 400 and 700 buffalo generations may have had time to elapse since the buildup of these environmental pressures.

The foregoing concerns the western buffaloes that may have retreated southward from the drying north, in response to the desiccation of the Saharan region. Subequatorial forest, which formed a barrier to a southward migration of grassland species, did not extend all the way to the eastern seaboard of the African continent. In the east, any ecological pressure caused by an inflow of animals from the northwest, as they retreated from the advancing desert, could be relieved by an overflow toward the south, into the high grassy plateau region that stretches from the great lakes to the southern tip of Africa. In the east, therefore, the desiccation of the north probably did not bring about any great ecological crisis. The buffaloes inhabiting regions between the Ethiopian highlands on the east and the forest on the west, at the top of the savanna "corridor" to the south, were never compelled to abandon an existence in large groups in grasslands, their preferred life pattern. The buffaloes of these regions are of the *aequinoctialis* variety, merging southward into the *capensis* or *radcliffei* varieties (the Cape buffalo). The rather minor differences between the Sudan and Uganda buffaloes of the *aequinoctialis* conformation and the Cape buffaloes occurring further to the south – mainly the larger size of the latter type – may be explained by the particularly favorable environment of the southern plateau regions and possibly the greater altitudes. In the gregarious *Syncerus*, the main obstacles to large group size are food and water restrictions. The very favorable conditions of the plateau led to larger groupings, besides having a direct effect on body size through high-quality nutrition. Group size may also be a factor in the increase of body size of individuals composing the group: The larger the group, the more bulls will compete for cows in order to mate. Before mating, bulls go through a process of elimination of potential sires by means that favor mass and strength (Chapter 8). The larger the number of competitors the greater the chance that exceptionally big individuals will be present. Thus the chances of cows mating with exceptionally big males increase with the size of the group to which all belong. In this way, the southern bush/grassland buffaloes, with their tendency to live in large groupings (Chapter 4), may be bigger not only because they are better fed but also because there is selective breeding that favors largest size.

In western Africa, following the influx of grassland animals from the drying-out northern regions, an ecological balance must have been approached once more. Under those conditions, accompanied by a drop in competition for food, a proportion of any buffalo population that may have been forced into the forest would be likely to move outward again, toward grasslands. They would have moved to, and outward from, the northern forest margin, obviously not back into the arid north but probably mainly eastward, along the remnant belt of grassland between the forest and the desert. The process may have been two-sided, with buffaloes drifting out of the forest, while the forest was still receding because of continued desiccation. These animals, owing to improved diet, greater exposure to sunlight, and increased intraspecific contacts leading to genetic diversification, would start to increase in size and become modified toward a grasslands form.

The *aequinoctialis* or a small-bodied variant of the *capensis/radcliffei* type may be, or may approach, the ancestral form; dense forest buffaloes of the Ituri type may be stunted forms of open country grass eaters forced into dense forest habitat; the western buffaloes of the bushcow type may be stunted forms reemergent into grasslands with increasing opportunities of genetic mixing; and the typical southern and southeastern Cape buffaloes may be modifications of the ancestral type, caused by especially favorable environmental conditions, that possibly also favor certain genetic trends, as suggested above.

The classification adopted here is of one species, *Syncerus caffer*, with two subspecies, *caffer* and *nanus*; the first subspecies includes two main varieties, *aequinoctialis* and *capensis/radcliffei*; the second subspecies includes at least two varieties, *nanus* (dense forest and similar forms) and *brachyceros*.

3

Methods of study

Field observations were made from motorcars, usually four-wheel-drive types, and aircraft, but mostly on foot. Observation from a motorcar was found to be least tiring and offered the advantage of a well-supplied mobile base. It was, however, possible only in open, relatively unbroken country, in the dry season, and in connection with buffaloes that tended to be stationary at the time. Following a moving herd in a motor car was usually unsatisfactory, as the buffaloes' behavior was clearly affected by the presence of the moving vehicle. Furthermore, buffaloes often traversed types of terrain that were not easily, or not at all, negotiable by motorcar. During rains terrain in which a motorcar could not be used was greatly enlarged: Many flatlands turned to swamp, dry-weather tracks turned to deep soft mud, and unbridged streams became major obstacles. The use of a motorcar for observation was unsatisfactory on moonless nights, unless a buffalo herd was quite stationary.

Observation from aircraft was a minor method. It was useful mainly for obtaining general views of herd patterns and counting.

Observation on foot was the most successful method (though physically demanding), as it was possible to maintain contact with the buffaloes 24 hours a day and to spend much of the time within the confines of the herd. I was able to remain in constant contact with the buffaloes for periods of up to 72 hours, after which weariness made me incapable of attentive observation. Under some conditions, my attention limit was less than 72 hours. It was found that two persons, if suitably qualified, could remain near or among buffaloes for long periods of time without disturbing them. Three or more persons almost invariably caused a disturbance. Very few truly competent assistants were found, and I did most of the meaningful work alone. Occasionally I used a bicycle to transport myself from camp to a distant buffalo herd.

While making observations on foot, the volume and weight of equipment had to be kept as low as possible to reduce fatigue and increase freedom of movement. This equipment varied somewhat from occasion to occasion but normally included binoculars, a 35-mm single-lens-reflex camera equipped with a standard and a 200-mm telephoto lens, compass, notebooks, maps, rations in the form of a complete powdered food, such as Complan (removed from its tin to reduce its weight), pressed dates, vitamin and salt tablets, and a canteen of water. In cold weather, a sweater and a light wrap, preferably a plastic thermal blanket, were necessary. During the rains, several plastic sheets, usually green garbage bags, were carried. A small cloth bag containing fine wood ash or silt was considered a necessity in order to keep track of air movements when near buffaloes. A gun was usually omitted because of its weight and unwieldiness, but in areas where there was a strong chance of meeting a wounded buffalo or where elephants tended to be aggressive, a gun was necessary. During 1977, an audio cassette recorder was usually included. The rest of the equipment included a flashlight, a tin for drinking and cooking, matches, nylon rope, knife, field dressing, spare film, cassettes, and batteries, and a few other small items. It might be noted that, throughout my years in the African bush, I normally drank unboiled and untreated water, from rivers, marshes, and waterholes, during all seasons, without ever contracting any significant malady as a result. This is not to be taken as a recommendation, however, and furthermore the taste and appearance of bush water may be foul. Nevertheless, if contact is to be maintained by an observer on foot, then water is the one item that cannot be carried in sufficient quantity.

The distances at which observations were made varied. They depended on one or both of the following factors: the desirable distance for the type of observation intended and whether the buffaloes could be approached, undisturbed, to that distance. If the main purpose was the study of a herd as a whole, or large parts of a herd, a distance of 60 to 80 meters was found to allow a good overall view of the activity, without missing details. Fortunately, this distance coincides with the common flight distance for most buffalo herds in the open, although if buffaloes are hunted from motorcars in open flatlands, their flight distance may be well over a kilometer, as for example on the Liuwa plain in western Zambia (R. A. Conant, personal communication, 1977). Remaining for many hours at a distance of 60 to 80 meters from a buffalo herd was not normally difficult, but certain precautions had to be taken for success. My position relative to the wind direction and the herd was crucial. Wind blowing steadily or in gusts directly from the observer toward the buffaloes caused the herd to move away and, even when it did not move far, all or part of the herd remained at the alert.

On the other hand, if I was fully exposed to view on the downwind side, it often caused many members of the herd to remain tensely at the alert, in an attempt to identify the intruder without the help of smell. It led to restlessness and obvious behavioral adjustments. Not infrequently, the tension built up to a point where one or several individuals displayed flight behavior and precipitated a general stampede or retreat. The remaining alternatives, however, often produced good results. One of these alternatives was to remain undetected by the buffaloes regardless of my cover. The other was to allow myself to be identified, by showing myself in a position from which the wind carried a light scent to the buffaloes, and have my innocent activities monitored by them, which appeared to give them a sense of safety. When I was accepted by the buffaloes at a given distance, it was most important to retain that distance. Even a slight decrease of it caused alarm as soon as noticed and started a buildup of tension that might take an hour to dissipate. This was true even on occasions when I established myself at distances in excess of the customary flight distance. Even then, a decrease of the established distance caused alarm. To become accepted by a buffalo herd at a distance that was less than their customary flight distance was possible but difficult. Unless the terrain was very flat and nearly devoid of shrubbery or trees, it was usually possible to remain undetected without losing the capability to observe.

If the purpose of observation was to be the study of individual behavior or interactions of only a few individuals, the most advantageous position for the observer was within the herd. It was difficult and often impossible to maneuver myself into such a position when working on herds that were heavily hunted but it was not difficult when dealing with buffaloes that led a fairly relaxed life. Here again two alternatives existed: Either to remain undetected by the herd or to be detected but tolerated by it. If the decision was to remain undetected while making observations, the most difficult part was to succeed in penetrating the herd without raising an alarm, because the likelihood of discovery was greatest when at a few meters or a few tens of meters from the flanking animals, which were watching for danger threatening from outside. If, however, I had succeeded in slipping through the herd's guard and reaching the center of the herd, I was among much less alert animals. Furthermore, the scent of the herd appeared to have a partly inhibiting effect on its members' olfactory sense or reactions. A further aid to remaining undiscovered, once in the midst of a buffalo herd, was the fact that either the buffalo's eyesight or its ability to identify an image is poor at very close quarters. Once among the buffaloes, immobility was essential, except for very carefully planned shifts in position. The most difficult condition to attain, while physically active in hot country, was to keep my own body odor at a very low level. Body odor, especially that trapped in clothing, promoted discovery by buffaloes, whereas if body odor could be kept down, not so much by ceasing to perspire as by wearing a minimum of clothing, the chance of discovery was much reduced. Dull-colored clothing, usually brown or khaki, was worn for the sake of inconspicuousness, and glittering objects, such as my wristwatch, were concealed, as they tend to draw the attention of wild animals.

The second close-study method, that is, exposing myself and being accepted by the herd as no threat, could be used only on buffaloes that had not developed a ruling fear of man owing to heavy and steady hunting. It consisted of finding a buffalo herd that grazed across, rather than into, the wind, waiting hidden ahead of it in its probable path, and exposing myself when the foremost animals were at a distance of 10 to 15 meters. The forward buffaloes invariably stampeded back, into the ones that followed. This sometimes caused a general stampede, leaving me devoid of any subjects for study. Often, however, the rear buffaloes would not stampede, causing the front animals to slow down, then stop and turn around to look toward me. At that stage, the foremost buffaloes were usually some 30 to 50 meters from me, the herd being in a congested mass. The buffaloes watched me attentively. I then performed a series of unhurried assorted movements that included walking but always parallel to the front of the herd, never toward it. This was continued for anywhere from a few minutes to 45 minutes. At some point, the foremost buffaloes either displayed flight behavior and caused general flight or one and then more of them started to graze, usually moving diagonally toward me; that is, deflecting but slightly from their original path. From then on, I was able to stay with the herd, provided I remained in view and did not make abrupt movements. In Zambia, this method was still practicable until 1974, in the Kafue National Park and its surroundings, as well as parts of the Zambezi Valley. In recent years, however, excessive hunting outside wildlife reserves and an increase of poaching inside them have made the surviving buffaloes more fearful than previously, and the method did not usually succeed.

In the case of large buffalo gatherings, of 600 or more individuals, it was sometimes judged best to make observations from as far away as a kilometer, in order to have an overall view of the group, and depending heavily on binoculars for details.

Tracing the movements of buffalo herds could often be done more efficiently by "spooring" than by closely following the buffaloes. The amount of definite information that can be obtained from buffalo spoor is quite large, and I was able to devote my whole attention to the subject in hand, without having to worry about such matters as discovery or influencing the animals' behavior by my presence. *Spooring* consists of three main components: The direct recognition of marks, such as footprints, body or muzzle marks, droppings, or marks on vegetation; the significance of marks, especially in combination with one another; and the construction of a time scale in order to make possible the placing of deduced events in time, by using such occurrences as the time and place of the last sighting of the herd, the time of a spoor-smudging rainshower, or particularly desiccating or moisture-preserving weather conditions, crushing of blooms at a stage of unfolding indicative of a particular time of day, and much else.

In order to study certain behavioral patterns, it is essential to be able to recognize individuals. This presented a problem that differed in scope from that facing the student of one particular group of animals, perhaps a family group of 10 or 20 individuals or a relatively small animal population scattered over a known home range. There were occasions on which the problem of individual recognition resembled the examples just given, for instance, while observing an isolated group of buffalo males or a remnant herd of only a few tens of individuals. Most often, however, I was faced at any one time with animals in the hundreds, which, moreover, occasionally interacted with other similarly large congregations. The configuration of animals in a buffalo herd blocks many individuals from view. In these conditions, I found it impossible to identify all the individuals within a herd sufficiently thoroughly to be able to recognize them whenever I met them, although usually I was able to follow the activity of a given individual throughout any one period of observation. Prolonged observation of some buffalo herds did, however, lead to thorough identification of a large proportion of the individuals, sufficiently large to permit, for example, recognition by quick inspection that two previously separate herds had joined. As appears to be the case with many other mammalian species, buffaloes show much individual variation, in terms of physical and behavioral features. After a degree of familiarization with a given buffalo, the probability of misidentifying it seems not much greater than the probability of misidentifying one's dog from among others of the same breed.

Field notes were taken as circumstances permitted, that is, when there was time to write them. Each note was numbered by the hour and date of the start of writing. For example, a note started at half past five in the afternoon of July 10, 1975 would have been coded "1730/07/10/75." Illustrative photography was attempted whenever time and light conditions permitted. Previous to 1977, vocalizations were carefully listened to and described to the best of my ability. In 1977, a lightweight cassette recorder was carried, and many vocalizations were recorded. Frequently, conflicts arose whether I should give priority to direct observation of an event, to photography, or to sound recording and, if direct observation, then how much watching time I should sacrifice to note taking. There is no formula answer nor any base for such a formula, and so depending on the circumstances I had to make snap decisions.

Often, the main area of observation had to be left to a last-minute decision. Before leaving for the field I made a tentative plan of work, but if later circumstances provided unusual opportunity to study a different behavioral pattern, the plan might be adjusted to take advantage of this opportunity. Conversely, if the original plan was, for example, to penetrate a herd and observe mother–calf interactions but the herd did not allow itself to be penetrated, then attention may have been shifted mainly, perhaps, to danger-detecting behavior of the peripheral animals. Techniques approaching controlled experiments were seldom possible. They were, however, occasionally resorted to – for example, when a herd was "spooked" on purpose to determine whether, and where, it would resume its normal route.

The programming of the observations may be summarized as follows: A field of behavior, such as all or some agonistic or ingestive behavior, was chosen as a main objective. Attempts were then made to observe instances of the target behavior. If it was later determined that adequate observation of this behavior was, for practical reasons, impossible at the time, a list of other priorities was consulted, and the herd was scanned for any other behavioral patterns and events that might be happening in it, to decide whether they merited more immediate attention. After several outings, consolidation of the field notes indicated whether data on a given behavioral topic were building up satisfactorily and whether the topic was, practically speaking, worth pursuing. Plans could then be revised accordingly. This was an effort to systematize the observations without unnecessarily losing hard-to-come-by data through excessive systematization and resultant exclusions.

For comparative purposes, some observations were made on the behavior of captive buffaloes, both captured in the wild and born in captivity. A literature search was made in the course of the study and the sources consulted. The search included work covering all aspects of the African buffalo and publications bearing on the behavior of the African buffalo are included in the References.

4

The herd

The basic herd

From 1964 to 1966, I observed a buffalo herd in southwestern Kenya. The herd stayed around Lolgorien, between the western margin of the Isuria Plateau and the Migori River (Figure 4.1). I lived at the edge of the buffaloes' home range between October 1964 and August 1965 and contacted them periodically at other times.

I noted that the majority of the 100-odd individuals comprising the herd tended to form three groups. Although adult males were present in these groups, adult females predominated. All the suckling calves and most older juveniles and subadults were included in the groups. These groups were sometimes separated from each other by distinct breaks, while at other times they tended to merge, and this applied regardless of whether the herd was on the move or stationary. A few animals, chiefly males, did not appear to favor any one of the groups but remained apart. When I had become sufficiently acquainted with the herd to be able to positively identify 55 adult individuals, that is, between 39 and 40 percent of the total number, or over 63 percent of the adults, I was able to conclude that even when the buffaloes were in a compact mass, the three groups did not mix with each other except around their edges and that the membership of the groups stayed largely unchanged during the period of my contacts with the herd.

Leisurely examination of the herd was possible when it rested or grazed at the base of the steep-sided Natakili Hill and in other places where I could observe them by climbing trees. The examinations tended to confirm that the three groups within the herd had permanence. I called them A, B, and C. Table 4.1 shows the composition of the herd as it was noted in January, July, and October 1965 when all the changes could be explained by movements of individuals that I was

41

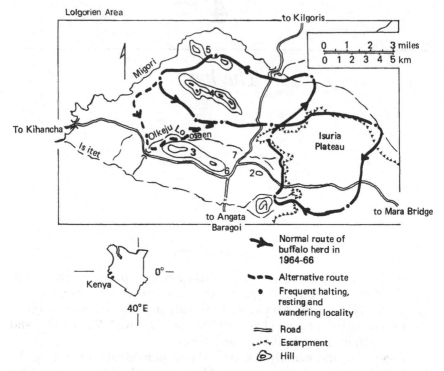

Figure 4.1. Map of the Lolgorien area, southwestern Kenya, showing the normal route followed by the resident buffalo herd in 1964–66. The area was inhabited by pastoral Maasai and frequented by hunters. West of the Isuria Plateau, woodland was common near hilltops and along watercourses, with grassland in most of the intervening localities. On the Isuria Plateau, both grassland and woodland (mainly open) were present. The buffaloes utilized all types of country, with a preference for grassland and using woodland as shelter from human threat. Key: (1) Natakili Hill; (2) Larumbas Hill; (3) Kimoigoyen Hill; (4) Olenkapune Hill; (5) Olorropil Hill; (6) Police post; (7) Author's mud huts.

able to identify. In Table 4.1, the individuals are subdivided according to age, sex, and group. There remained doubts with respect to some individuals as to which group they should be assigned. In such cases one of two courses was taken: Either the individual was assigned, on the basis of its movements and locations within the herd, to the group deemed most likely to be the correct one, or, if that appeared to be an unreasonably arbitrary procedure, the animal was listed as not belonging to any of the three groups.

Among the adult individuals, I learned to identify 10 females and 4 males in group A; 10 females and 6 males in group B; 6 females and 3 males in group C; and 2 females and 14 males that did not appear to be

Table 4.1. *Breakdown of the Lolgorien buffalo herd by age, sex, and apparent allegience to groups*

Group	Females				Males				Total
	Adult	<1 yr	1–2 yr	Subadult	Adult	<1 yr	1–2 yr	Subadult	
January 1965									
A	20	3	2	4	4	4	2	4	43
B	15	3	1	4	8	3	2	3	39
C	13	2	2	3	5	3	1	3	32
Other	4	0	0	0	17	0	0	6	27
Total	52	8	5	11	34	10	5	16	141
July 1965									
A	18	5	2	6	4	3	3	3	44
B	15	5	2	5	5	4	2	1	39
C	11	3	1	3	6	4	1	1	30
Other	3	0	0	0	15	0	0	6	24
Total	47	13	5	14	30	11	6	11	137
October 1965									
A	18	7	3	5	5	5	4	2	49
B	16	5	1	3	5	4	2	1	37
C	12	4	2	1	3	4	0	1	27
Other	2	0	0	0	13	0	0	5	20
Total	48	16	6	9	26	13	6	9	133

The following percentages are derived from the above
(total numbers in parentheses):

	January 1965		July 1965		October 1965	
All adults	70.0	(86)	56.2	(77)	55.6	(74)
All females	53.9	(76)	57.7	(79)	59.4	(79)
All adult females	36.9	(52)	34.3	(47)	36.1	(48)
All males	46.1	(65)	42.3	(58)	40.6	(54)
All adult males	24.1	(34)	21.9	(30)	19.5	(26)
All juveniles and subadults	30.0	(55)	43.8	(60)	44.4	(59)

permanent members of any of the three groups. This made it possible to recognize some changes that occurred, shown in Table 4.2 and Figure 4.2.

In July 1965, 7 individuals were observed to have relocated from one to another grouping within the herd with respect to January 1965, and in October 1965, 5 individuals were observed to have done so with respect to July 1965. This gives an average of 1.2 individuals relocated per month during the 6-month period from January to July and 1.7 per month during the 3 months from July to October. During the same two time intervals, 9 and 5 individuals, respectively, disappeared from the herd. Thus by July the adult population sample of 55 identifiable indi-

Table 4.2. *Relocation of recognizable individuals within the several groupings forming the Lolgorien herd*

	January–July 1965		July–October 1965	
	Losses	Gains	Losses	Gains
Females				
Group A	−2 (XX)	+0	−0	+0
Group B	−0	+0	−0	+1 (reclas.subad.[a])
Group C	−2 (OX)	+0	−0	+1(O)
Other	−2 (XX)	+1 (C)	−2 (CX)	+1 (reclas.subad. from C)
Males				
Group A	−0	+0	−0	+1 (O)
Group B	−3 (OXX)	+0	−1 (O)	+1 (O)
Group C	−2 (OO)	+3 (OOO)	−3 (OXX)	+0
Other	−5 (CCCXX)	+3 (BCC)	−4 (ABXX)	+2 (BC)
Sexes combined				
Group A	−2 (XX)	+0	−0	+1 (O)
Group B	−3 (OXX)	+0	−1 (O)	+2 (O reclas. subad.)
Group C	−4 (OOOX)	+3 (OOO)	−1 (O)	+1 (O)
Other	−7 (CCCXXXX)	−4 (BCCC)	−6 (ABCXXX)	−3 (BC reclas. subad. from C)

Symbols used: The number of individuals by which a given group has decreased is preceded by a minus sign, and the number of individuals by which a given group has increased is preceded by a plus. The parentheses contain one symbol for each individual animal displaced and, in the "Losses" column, indicate where an animal went, whereas in the "Gains" column they indicate whence it came. Of these bracketed symbols, A, B, C indicate the three groups, O indicates individuals from the "Other" category (as in Table 4.1), and X indicates individuals which have disappeared, i.e., left the herd or met death in any way. For example, an entry that reads
Group A   −2 (BX)   +1 (O)
means that two individuals left group A, one of them transferring to group B, the other leaving the herd altogether, while one animal from the Other category joined group A. See also Figure 4.2.
[a] Indicates individuals reclassified to adult status.

viduals, or 64 percent of the January adult count, was reduced by 9 to 46, or 56 percent of the July adult count. It was further reduced, by 5, by October, to 41, that is, 56 percent of the October adult count. Thus the sample used for the purposes of these observations, that is, 55 adults gradually decreasing to 41, was never less than 56 percent of the total adult population. If it is assumed that the observations made on the sample apply in simple proportion to the total adult population, then the average relocations within the herd become about 1.9 per month in

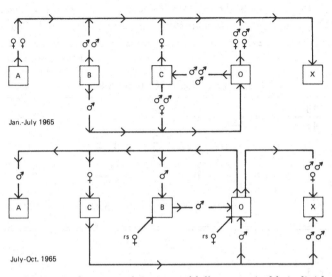

Figure 4.2. Diagram showing relocations of fully recognizable individuals within the several groupings forming the Lolgorien herd. Groups A, B, and C are denoted by their respective letters, the Other category (as in Table 4.1) by O, and individuals that have disappeared, i.e., left the herd or met death in any way, by X. Subadults reclassified as adults are denoted by rs. Each male and female symbol denotes one individual. The arrows point *away* from the groupings losing the individuals shown and *toward* these individuals' destinations.

January–July and about 2.8 per month in July–October. If we try to estimate the number of relocations per year by using a single weighted average based on these two figures, that is, 2.2 per month, we obtain the annual relocation estimate of 26.4 adult individuals. If these estimates are of the right order, they indicate considerable stability of the groupings within the herd. The majority of the individuals marked X in Table 4.2 were removed by deaths that I was able to verify.

During the period from January to October 1965 the Lolgorien herd decreased in numbers from 141 to 133, or by 5.7 percent. During the same period the adult population decreased by 14 percent, and the adult males by 23.6 percent (working from Table 4.1). These figures probably reflect increased hunting. At the start of the observations the herd was under little hunting pressure because the bridge across the Mara River on the Narok Road had been destroyed by previous floods, and access from Nairobi and adjacent settled areas was difficult (the local inhabitants, mainly Maasai, did not hunt buffalo). By October 1965 hunting by outsiders had again increased. A tendency for the proportion of adults within the herd to remain relatively constant during January–October 1965 is indicated by the ratios of certain classes of individuals to the total numbers in the herd (Table 4.3), but this may be fortuitous.

         The African buffalo

Table 4.3. *Proportions of different age categories in the Lolgorien herd and their percentage variations between January and October 1965*

Ratio	January 1965	July 1965	October 1965	Maximum percentage variation in ratio
Calves[a] $\dfrac{18}{\text{Total herd } 141} = 0.128$		$\dfrac{24}{137} = 0.175$	$\dfrac{29}{133} = 0.218$	41.3
Juveniles[b] $\dfrac{10}{\text{Total herd } 141} = 0.071$		$\dfrac{11}{137} = 0.080$	$\dfrac{12}{133} = 0.090$	21.1
Subadults[c] $\dfrac{27}{\text{Total herd } 141} = 0.191$		$\dfrac{25}{137} = 0.182$	$\dfrac{18}{133} = 0.135$	29.3
Adults $\dfrac{86}{\text{Total herd } 141} = 0.610$		$\dfrac{77}{137} = 0.562$	$\dfrac{74}{133} = 0.556$	8.9

[a] 0–1 year old.
[b] 1–2 years old.
[c] 2–4 years old.

It must be made clear that individuals from various groups did occasionally drift out of their own group, singly, in pairs, or even in threes or fours, but they later returned, and such absenteeism involved only a small part of a group's membership at any one time.

During the same period I noted that some adult females seemed more socially prominent than others. Whenever anything at all remarkable was noted, it seemed that one of these females was somehow involved. I began to look for any pattern that might confirm whether this observation had validity. I noted that the socially prominent females definitely spent more time near the front and periphery of their group than did other females, and that when they moved they were usually followed by other females or by subadults and juveniles. Thus within the moving herd if one of these females swung a few degrees to the main direction of herd movement, some other animals would do the same. Further, I noted several instances of subadult males making way for these females, a behavior not seen with respect to the remainder of the females. But I also noted that on the occasions when they drifted out of their own group these females did not appear to influence the activities of animals that belonged to other groups. All the socially prominent females were fully mature, and all the aging females in groups A, B, and C belonged to this behavioral type. While a major move on the part of one of the socially prominent females normally led to at least the beginning of a similar move by several other individuals, adult males were not so emulated, although females and nonadults of a group made way for them.

These observations appeared suggestive of a hierarchy. Some of the evidence suggested that this hierarchy did not transgress the boundaries of the groups that appeared to compose the herd. On this I based a supposition that some or all African buffalo populations are composed of social units smaller than the herd.

Subsequently it was found that this pattern could be detected in other herds besides the Lolgorien herd and that it was a common one. The degree to which these groups are imiscible with the rest of the herd, as well as their separate hierarchies and their degrees of independence, were subsequently observed. It was found that these groups could and did act independently of the herd to which they belonged at a given time. Each group had a relatively simple hierarchy of its own.

Whenever a herd fragmented, it always did so with the component groups essentially intact. These groups are not unlike family groups in certain other species, although they tend to be more inclusive than family groups. They appear to be the smallest self-contained social units of *Syncerus* and they vary in size from a very few animals, particularly in forest races, to possibly 100 in races of *S. c. caffer*, though most of them contain between 30 and 60 individuals. I refer to these groups as *basic herds*.

The basic herd is the domain of the female. As a rule, she does not for long stray away from it. She is the essential structural part of it, while the adult male only associates with the basic herd as season and inclination dictate, although many males adhere for long periods to the same basic herds. Once part of the basic herd, the male has active roles to play, but these roles do not shape the basic herd. He fits into a niche that is created for him by the others, even though it is not subordinate. There is something similar in the position of the buffalo bull in the basic herd, having high status but not supremacy, to the constitutional monarch in his kingdom, with the difference that in a basic herd several bulls are tolerated simultaneously. In the dominance scheme of *S. c. caffer*, any adult male appears to be dominant to any female, with the special exception of a certain type of barren female (see Chapter 7). The functions of the male and the female in the buffalo society do not overlap in such a way as to make this male dominance an usurpative factor, and on the basic herd level there exists a female hierarchy which is more effective and meaningful to the life of the basic herd than the ultimate hierarchy which places the male on top. The management of the young rests with the females. This includes feeding, herding, guiding to safety, defending, reassuring or stroking, warming, playing. I have observed only two situations in which an adult male may take the initiative to promote the welfare of the juveniles: defense and play. In all other situations the bull is passive, even sometimes aggressive, toward juveniles.

In any one herd, there are clear physical differences between some individual buffaloes, while strong physical similarities exist between some others. Such evidence as may be available suggests a direct blood kinship between the animals resembling each other. It is not difficult to spot groups of females, habitually keeping company with one another within the herd, that are much alike but vary in age. I consider on circumstantial evidence that such females are directly related.

I noted in the Lolgorien herd, and later in other buffalo herds, that where females could be classified into several distinct physical types, individuals belonging to one of these types tended to show dominant behavior toward individuals belonging to some other physical type when the interacting animals were of the same age. This was emphasized further by a number of cases where an older individual of the subordinate physical type showed subordinate behavior toward an obviously younger animal of the dominant type. Assuming that the tendency for individuals of the same physical type but different ages to stay together has meaning, and that close physical resemblance is the result of close heredity, it appears highly probable that dominance status among buffalo females is often passed on from mother to daughter. Thus kinship groupings in females appear to be the foundation of basic herds.

In short, the typical herd of *S. c. caffer* is composed of smaller groups, defined as basic herds, whose membership tends to be constant, which comprise individuals of both sexes and all ages, and which have internal hierarchies of their own. When a herd falls apart, these groups remain intact. They are the smallest fully socially and reproductively viable groupings of *S. c. caffer*.

Group size

In most localities where *S. c. caffer* occurs today, its numbers have been so reduced that group sizes are limited by the smallness of the total regional population, before any other factor can come into play. In the several regions which still contain buffalo populations so large that the above does not apply, the smallest separate groups are formed by bachelor males. Bachelor males occur most commonly in numbers of one to a dozen. Ansell (1960) mentions the occasional occurrence of bachelor groups numbering up to 30 in Zambia. Selous (1881), commenting on the Chobe River area as it was in the 1870s considers a bachelor group of 15 unusually large; it would appear it was the largest that he had ever encountered. Sinclair (1977) gives 51 as the largest bachelor group recorded in the Serengeti. The largest such group ever observed by me contained 38 individuals and was seen on the Busanga flats, Zambia in October 1971. It may be safely said that bachelor groups in excess of 20 are rare.

Given a large population of *S. c. caffer*, mixed groups are normally much bigger than any bachelor groups. But an apparently innate tendency of buffaloes to gather in large companies is offset by the limitation imposed on the number of animals in a group by food and water supply as well as the inability to maintain social cohesion above certain group sizes. Disruptive potentials set in when groups number thousands, or under mediocre range conditions even only hundreds, of individuals. These disruptive potentials are activated when there is a decrease in food or water supply or an increase of external stresses requiring quick responses by all group members, for example, predation by lion or man.

The limitations of social organization on the group size may be fairly constant for the species, since it largely depends on innate biological characteristics, such as the development and functioning of the nervous system. One would expect this to be uniform throughout the species and to lead to a standard complexity of organization within the groups. The alimentary factor in the upper limitation of group size is very variable, on the other hand, being governed by the seasons, cyclical food supply, and change in habitat, especially cover.

When a buffalo gathering exceeds about 600 individuals, a weakening of social cohesion becomes apparent and tendencies to divide into groups become more pronounced. The largest buffalo gatherings ever reported are of 2,000 to 3,000 individuals, and there are indications that it is not practical for *Syncerus* ever to form larger groups, because of the grazing efficiency of smaller groups.

A gathering exceeding 600 occupies too large an area, especially when grazing, and involves too many individuals for efficient transmission of signals. Another factor tending to make very large grazing groups less efficient is the drive of dominant individuals toward maintaining peripheral positions in the group while grazing. Groupings situated centrally with respect to a gathering as a whole were occasionally observed marching through the scattered gathering of grazing buffaloes in the wake of their leaders, who apparently wished to secure a position nearer to the edge. This increases the amount of movement in such centrally situated groupings and thereby decreases grazing. When this behavior is repeated by more than one group within a gathering it tends to produce either extended lines or columns of individuals – which is an inefficient deployment for large numbers of buffaloes when grazing as those in the rear go over ground already grazed and trampled by those ahead – or a laterally very widespread deployment, which tends to cause the large gathering to disintegrate into several smaller ones, likely then to drift apart.

For large groups of buffaloes to remain stable over long periods, the requirements are: plentiful graze over a vast uninterrupted area, plentiful water supplies throughout that area, a low level of interference by external agencies, and a tendency for any interference that may occur to have a pattern that can be responded to in a routine manner. Several of these conditions are best met in extensive floodplains and broad river valleys with much lateral drainage. It is indicated in old records (e.g., Livingstone, 1857, 1865; Selous, 1881, 1908), as well as from more recent observations, that the largest single buffalo gatherings in the past did, as they still do, occur in such localities.

An important element in a discussion of buffalo group size is the late-nineteenth-century occurrence of a catastrophic epizootic, the rinderpest (caused by the virus *Tortor bovis*), which started in the Red Sea region around 1890 and spread southward, reaching the southern tip of Africa by the end of 1896 (Cloudsley-Thompson, 1967). This macabre episode, described in some detail by a number of authors, decimated many wild ungulate species, as well as domestic stock, in the eastern and southeastern portions of Africa, and was particularly hard on the buffalo, which was wiped out in several regions. Rinderpest continued to make itself felt among buffalo populations, though with generally diminishing intensity, well into this century (Uganda Game Dept. Annu. Repts 1925–1945; Cloudsley-Thompson, 1967; Sinclair, 1977). In recent decades, probably largely owing to the diminution of rinderpest, buffalo populations have greatly increased in regions which are either protected or inaccessible to hunters, and there are indications that group sizes in these areas have also correspondingly increased.

A second important element in considering buffalo group size is the progressive elimination of buffaloes through hunting as a consequence of expanding human occupation of Africa. Just when this phenomenon became of great importance is arguable, in the same way that the year after which rinderpest is relegated to a position of minor importance must be arbitrary. These two phenomena, however, tend to overlap and mask one another. Moreover, both are expressions of the recent spread of *Homo sapiens* and may be considered to have affected the buffalo populations for unprecedented reasons.

At this point it seems worthwhile to diverge a little further and recall that over the past several thousand years, since the beginning of the current interglacial stage, the African climatic scene has been undergoing a change (e.g., Clark, 1960; van Zinderen Bakker, 1962), which has been dramatically reflected in changes of vegetation and surface water distribution.

All these factors have a bearing on buffalo group size. It is amid such considerations that one begins to wonder in what measure any observations of social patterns may be generalized without introducing serious inaccuracies.

The herd defined

The smallest grouping of *S. c. caffer* with social and reproductive viability, the entity which I call the basic herd, normally forms part of a larger grouping. This larger grouping is composed of basic herds, and in addition contains up to three types of specialized groups: bachelor, invalid, and juvenile clubs. Thus in a population of *S. c. caffer* there occur different types of groups, some of which may occasionally exist separately from others, and therefore the noun *herd* lacks clarity if used without a strict definition.[1]

I use *herd*, when applied to African buffaloes, to indicate a biological group made up of a few to a few hundred individuals of both sexes, which functions on its own for weeks, months, or occasionally years at a time and acts as a social and reproductive entity throughout that time. The most important point of this definition is that it excludes some specialized groups, such as isolated bachelor or invalid clubs and the largest type of African buffalo gatherings known, attaining over 2,000 individuals, which tend to lack social cohesion. The definition does include, of course, a single basic herd operating on its own, rather than as part of a larger group.

The herd is the most common grouping of African buffaloes; it is the normal grouping. There appear to exist several hierarchies among the African buffaloes, some of which operate within basic herds and some of which result in binding a number of the latter together into a herd.

Herd size

The buffalo herd varies in size depending on the quantity and distribution of food and water, as well as physiographic, vegetational, and predational factors. Within the eastern, east-central, and southeastern regions of Africa, herd sizes, excluding remnant herds, vary from fewer than 20 (my lowest exact count was 19, within the Busanga buffalo population in Zambia) to approximately 600 individuals. But gatherings in excess of 500 are in my experience usually composite herds, that is, two or more herds that have seasonally drifted together. Just when a composite herd should be simply referred to as a herd is not always obvious and can only be decided by the degree of social cohesion between the herds that have come together. Sinclair's (1977) comments on the Grumeti River buffalo population in Tanzania's Serengeti National Park indicate that various censuses taken of the Gru-

[1] The *Oxford International Dictionary of the English Language* (Unabridged, 1957) gives, as the second of three definitions: "2. A company of animals of any kind feeding or travelling together; a school (of whales, etc.) ME." There is no entry of *herd* in *A Dictionary of Biology*, Abercrombie et al. (1976 reprint).

meti herd from 1966 to 1973 showed variations of up to a quarter of the largest number recorded, 1,750, down to 1,300 individuals. This variation of 450 individuals is, using Sinclair's data, 9 times greater than the smallest herds seen in the Serengeti, of 50 individuals, and exceeds by more than a quarter the average herd size in that region (350 individuals). These data are suggestive of fragmentation occurring in the buffalo gathering in question. Sinclair also states that many of the smaller groups in the Serengeti (from his observed minimum of 50 individuals to, I assume, as many as 200?) were almost certainly "splinters" of the herd that had temporarily diverged. These observations generally agree with my own in other regions, although I would interpret the smaller "splinters" to be the stabler entities (basic herds) and would expect the large Grumeti herd of over a thousand to be a gathering or a composite herd.

There are fairly numerous references to buffalo herds in the literature, describing them as "large," "very large," "small," and so on. This kind of observation is obviously of limited value. There are not many references to actual numbers, and a sizable proportion of those that exist gives the impression of very rough approximations. But a few figures have been published and are presented below, starting with the most recent and ending with the oldest. The word *herd* retains any meaning the original author may have given it.

For the Serengeti area Sinclair (1977, p. 120) states that herds range in size from 50 to more than 1,500 (May 1968) and cites one count of 1,750. He also states (1977, pp. 119, 121) that herds on Mount Meru (northern Tanzania) are of about 50 individuals and that a herd in the Ngurdoto Crater, between Arusha and Moshi in northern Tanzania, was estimated at about 300. In 1974 I photographed and counted 348 in the Ngurdoto herd.

For the Chobe National Park, northern Botswana, Sheppe and Haas (1976) mention herds of 300 at the start of the rainy season (October or November) and one of over 1,000 at the end of the rains (in March). But a rainy season census in the same area, reported by the same authors, gave a maximum herd size of only 23. For the Wankie National Park, northwestern Zimbabwe (then Rhodesia), the following seven herd sightings were reported by Wilson (1975): over 600 (Chingahobi, June 28, 1968), about 1,200 (Mahoma Loop, July 15, 1972), 965 (Kennedy II, September 8, 1972), 814 (Ngweshla, September 7, 1973), over 800 (aerial count, Dolilo Springs, September 12, 1973), over 600 (Lipande, September 25, 1973), 700 to 800 (Ngamo, October 11, 1973).

For Ethiopia, Bolton (1973) cites reports of a herd 250 to 300 strong on the Mwi River and of one numbering about 50 in the Mago Valley.

For the Kruger Park, South Africa, an accurate aerial census of the Crocodile Bridge buffalo population, taken in July 1966, indicated seven herds numbering 454, 384, 670, 158, 172, 91, and 54 (Pienaar, 1969a, pp. 45, 46). In addition, the same source cites five more Kruger Park counts: 97 (Shabin herd, Pretoriuskop), 204 (large herd, Makamba, Lebombo flats), 91 (small herd, Bob, Lebombo flats), 116 (portion of Sabi-Sand herd), and 527 (large herd, Saliji).

For East Africa, Grimsdell (1969) reports four groups in the Queen Elizabeth National Park, Uganda, which varied between the following extremes over a period of 13 months: 122 to 138, 97 to 111, 94 to 108, and 23 to 25; and Foster and Coe (1968) mention a group of (presumably) 17 buffaloes observed in the Nairobi National Park in 1966. I observed the same group in 1966. These animals were introduced and resembled a remnant herd.

For the Congo (Zaire) basin low-forest areas, Sidney (1965) gives an average of about 20 for herds. In the Save river area of Mozambique (Safarilandia hunting concessions) buffalo herds of over 40 were reported (Dalquest, 1965).

For the Kagera National Park in Ruanda a herd of at least 264 was cited (Curry-Lindhal, 1961), and Verheyen (1954), using data derived from Belgian Congo's (presently Zaire) Albert National Park (northern sector), and the Upemba National Park, mentions herds of 50 to 200 as "very large." He states that the herd is never stable in its numerical composition. The same author (1951) mentions for the Upemba National Park that herds of about 40 occurred at high altitudes (1,400 to 1,800 meters) and that the average was 6 to 10 but that herds of 50 to 200 occurred.

Pitman (1934) refers to herds in western Northern Rhodesia (now Zambia) numbering 120 as "huge herds." This is the general area in which composite herds of around 2,000 have been observed in the 1970s (could a zero have been ommitted in Pitman's figure?).

Again in East Africa, Rockwell (1934) counted 250 in a Tana River area herd in Kenya, and in about the same area, by the confluence of the Tana and Theba rivers, Akeley (1921) saw a buffalo herd of about 500.

With reference to the forest-living remnant population of *S. c. caffer* in the Addo Forest, Port Elizabeth, South Africa, Fitzsimons (1920) cites about 15 as the usual herd size.

Selous (1899), writing about southeastern Africa before the rinderpest epizootic, thought of herds numbering 200 or 300 as "large." He placed most herds in the 50-to-300 range and referred to one herd of over 1,000 by the Chobe River, which he considered to be several herds joined together (i.e., a composite herd). A buffalo herd of 81 in the Libonda (Kalabo) area of Zambia is mentioned by Livingstone (1857), but from his other observations it can be inferred that most herds he saw in the region were much larger and that buffaloes were numerous.

The following interesting figures were gleaned in the literature on West Africa, where the tendency is to think of buffalo groupings as consisting of a very few individuals: Bourlière et al. (1974), discussing the Lamto area, Ivory Coast, cite native hunters' reports that within living memory buffalo groups numbered 50 to 60 individuals. Discussing the buffaloes of the Yankeri Game Reserve in northeastern Nigeria, Henshaw and Greeling (1973) mention herds of up to about 140. And Basilio (1962) writes that in (former) Spanish Guinea, in areas less frequented by man, buffalo herds attained 100.

To judge from this literature, buffalo groupings in the eastern, east-central, and southern portions of Africa tend to range from about 40 to about 1,500. There is wide agreement that groups of between 50 and 700 are normal, and nearly as much agreement that the limit of the normal range might be 1,000. There are a few reliable observations of groupings between 1,000 and 1,500, and evidence exists that larger ones occur. There are also a few observations suggesting that eastern buffaloes do on occasion herd in numbers considerably fewer than 50. Estimates from 100 to 300 are the most common.

The following are some of my own counts, in which the numbers refer to entities called herds under my definition of the term, already given, with composite-herd numbers added: Lolgorien, southwestern Kenya (1965), 133 to 141; Chenza Ridges, southeastern Kenya (1967), 42; Mabalauta area, southeastern Zimbabwe (1972), 350 to 450, 178 to 225, and 120; Mara River vicinity, Kenya (1964) 900 to 1,200 (composite herd?); Ruaha National Park, southern Tanzania (1972) 88, 260, 324, 412 (all from my aerial photographs); Luangwa Valley, eastern Zambia (1975) 300 to 600 approximately, combining to form groups of over 1,000; Zambezi Valley, south (Zimbabwe) bank, Mana Pools and hunting blocks area (1971), 80 to 162; Zambezi Valley, north (Zambian) bank, Feira to Chirundu (1971), 31, 61; Busanga area, western Zambia (1973), 311, 348, and composite herd of 1,950 to 2,200; same area (1974), 180, 250, 300, 370, 380, 450, and a composite herd of about 1,250; same area (1977) 19, 36, 53, 112, 140, 160, 220, 400, and a composite herd of about 850 (most Busanga counts are rounded off to the nearest 10); Lukwakwa area, western Zambia (1973), 61: Ngurdoto Crater, Tanzania (1974), 348 (photograph).

My own experience, therefore, tends to confirm the observations regarding group size recorded in the literature, with the following comments: Groups below 100 may be remnant populations, and groups below 50 are usually so. In other words, it is seldom necessary for buffaloes to exist in groups of these sizes because of limitations due to food or water. Groups above 500 are frequently composite herds which have gathered for as long as food and water supplies permit.

Seasonal changes in herd size

The relationship between herd size and seasonal changes in the habitat is no longer easy to observe in full, since very few localities remain where buffaloes are numerous enough to retain the potential to form very large herds and are sufficiently free and undisturbed to do so.

In Zambia, the Busanga region in the northern sector of the Kafue National Park is such a locality. The extensive Nanzhila composite herd in the southern sector of the same national park and the large buffalo concentrations in the Luangwa Valley of eastern Zambia are others. Localities from the Zambezi Valley southward recently known to contain large buffalo populations are within or near zones of recent or current armed conflict, accompanied by an inflow of humans equipped with automatic firearms. Their standing as game areas is jeopardized, with the current exception of the Kruger National Park in the Republic of South Africa and possibly the Chobe National Park in northern Botswana.

The best study area in east-central Africa is, in my opinion, the Busanga region where conditions are little altered. The Busanga buffalo population has not been seriously impeded by human agency in recent times, in the form of excessive hunting, settlement, or other development, including major tourism. The habitat is most favorable and the buffalo population adequately large for research purposes. It numbered no fewer than 3,000 in the main part of the area up to January 1978. This main part covers some 5,625 square kilometers and is located within a protected area of much greater size which encompasses the whole of the Kafue National Park, of more than 22,000 square kilometers, exclusive of adjacent game management blocks. In all, the maximum protected area may amount to 77,699 square kilometres, depending on the effective protection within the game management blocks (poaching was very heavy in many areas other than the Kafue National Park itself during the later 1970s). The movements of buffaloes in the Busanga region were usually not restricted or modified by physiography, human activities, or settlement. The Busanga flats belong to the environmental ecotype that harbored great buffalo populations of the past including, among others, the Kafue flats, parts of the Zambezi Valley, the Lake Rukwa area, and the Save (Sabi), Chobe, and Pungwe River valleys.

In East Africa, the most important region retaining the potential to form very large buffalo herds is undoubtedly the Serengeti National Park and, at least until recent years, its neighborhood both on the Tanzania and the Kenya sides. The majority of my observations in that general region were near Lolgorien, along the Mara River, and in the Loita Hills area between Maji Moto and the Mburobudi Hills, all in

Kenya and mainly situated just north of the localities where herding on a very large scale may occur at present. My observations in the Serengeti proper are not sufficiently numerous to be in the same class, statistically, as those presented by Sinclair (1977).

The largest group that I encountered just outside the Serengeti was in February 1965 some 3.5 kilometers east of the Mara River and 40 kilometers north of the Tanzania border. It numbered no fewer than 900 and probably not more than 1,200. It was in grassland with minor tree growth, and remained there throughout the day. The largest herd size cited by Sinclair (1977) for the Serengeti is 1,750, the average being 350. In the Busanga region, the largest group that I observed was between 1,950 and 2,200 (April 1973). The average herd size based on 19 counts over 1973, 1974, and 1977 was 416 and ranged from 19 to 2,075[2] (the last figure is the mean of the April 1973 estimates). An aerial count of four buffalo herds in the Ruaha National Park, southern Tanzania, in July 1972 produced an average of 271 and ranged from 88 to 412. An estimate of three buffalo groups made by me in October 1972 in the Mabalauta area of the Gona-Re-Zhou Game Reserve, Rhodesia (now Zimbabwe), produced an average of 265, with a range of 120 to 450. The July 1966 census of the Crocodile Bridge buffalo population in the Kruger National Park reported by Pienaar (1969) gives an average group size of 283 (seven herds ranging from 54 to 670). He also cites numbers for another five herds in the Kruger Park, ranging from 91 to 527 and averaging 207 (the two Kruger averages when combined give 245). The grand average based on the above figures for the five localities – Serengeti, Ruaha, Busanga, Mabalauta, and Kruger – is 309, ranging from 19 to about 2,000.[3]

Although the average herd sizes for the five regions correspond reasonably well, they cannot be presented without further comment. This is made necessary by the transfers that occur between those buffalo groups which together constitute one larger population. The following quotation from my 1972 summary notes on the Mabalauta area bears this out:

> On 10 October, the Buffalo Bend and the Mafuku herds came close to each other in the area of the Manyanda pan and some redistribution took place. Several small groups formed while both large herds diminished, the Buffalo Bend herd to some 350 individuals and the Mafuku herd to some 180 individuals. The Mafuku herd numbered 178 individuals on 19 October. On 21 October, the Buffalo Bend herd numbered at least 400 individuals. Observations suggest that the three herds observed to move within the Mabalauta area in October, 1972 are miscible to some extent.

[2] For the purpose of these averages, composite herds are treated as ordinary herds.
[3] See footnote 2.

Similarly Sinclair (1977) indicates graphically that in 26 censuses taken in the northern part of the Serengeti National Park, between July 1967 and September 1969, that is, averaging one census per month, mean herd size always differs by 16.7 percent or more between consecutive censuses, and the maximum difference between consecutive censuses is 92.6 percent.[4] The largest recorded mean herd size is 590 animals and the average monthly variation of mean herd size is 106.8 animals. The greatest difference in mean herd size during the 26-month period was 530, that is between 60 and 590. Furthermore, in July 1967 mean herd size was 120, in July 1968 it was 245, and in July 1969 it was 170. Thus Sinclair's graph, besides clearly indicating a seasonal increase (wet) and decrease (dry) in mean herd size, also indicated that considerable fluctuations in herd size are normal throughout the year in the Serengeti. The considerable month-to-month variation in mean herd size appears to be independent of the seasons to a large extent. This suggests that individuals or groups change herd allegience rather frequently.

Effects of water and pasture on herd size

A buffalo herd's daily regimen depends first of all on the distribution of water throughout its habitat. If water is to be found only in one or very few localities of the herd's home range, regular movements are necessary between pasture and water. On the other hand, if water is widely distributed throughout the herd's home range, the need for regularly spaced, and sometimes long, marches does not exist, and the daily behavior will be less regular and not diurnally cyclical.

These two types of water distribution demand from the buffalo different degrees of social conformity and organization. Making a frequent and regular move to water implies a synchronicity of digestive processes, so that the requirement for water will occur at about the same time in all herd members. It necessitates the resolution of such problems as: where to seek water, how to get there safely, and when to start moving toward it. This implies either some leadership functions or a level of allelomimetic behavior that are not necessary when pasture and water occur together.

The need to trek to water shortens the time available for other activities, paramount of which are feeding and rumination. The larger the herd, the more pasture is required in any one locality, and therefore the farther from water the herd may have to move. Thus the need to commute between pasture and water tends to work against the formation and persistence of very large herds.

[4] See Sinclair (1977, Fig. 47). The values I derived from this figure may be subject to minor corrections, as I obtained numbers by interpolation on the graph with the aid of a linear scale. This small amount of error (if any) is immaterial to the argument.

When there is no need to move from pasture to a watering place, that is, when water becomes available throughout, much of herd discipline becomes unnecessary. The herd is then seen to lose much of its shape. A grass-bearing area that is well watered tends to attract and retain much of the regional buffalo population, resulting in an increase in herd size, until such time as water or pasture conditions deteriorate. Pasture can deteriorate either because of a drying out or an excess of water. In the case of an onset of drought, the buffalo population becomes reorganized to fit a regimen based on regular commuting between distant pasture and water. Swamping out of pasture occurs when the latter grows on the floodplain of a river or a lake, such as the Kafue and the Busanga flats in Zambia or the margins of Lake Rukwa in Tanzania. As already mentioned, such localities are of importance in the study of *S. caffer*, as they appear to be a favorite part of its habitat.

During much of the year the vast open grassland of the floodplain offers the buffalo a nearly ideal balance of good pasture and water. As the rainy season advances, the water level rises in the middle of the plain and water spreads outward toward the margins, with the buffaloes moving before it. The rims of the floodplain are likely to abound in waterholes and small watercourses during the rainy season and to contain good pasture. Pasture is somewhat less continuous on the rims, owing to the interference of woodland, and more intensive search for suitable grass patches is required on the part of the buffalo, but, most importantly, vacating the open grasslands at such a time does not bring about a separation of pasture and water supply which would necessitate daily trekking. This is probably a real factor in the success of buffaloes in such habitats.

As the buffalo population moves outward and spreads around the rim of the floodplain, two factors begin to have impact on herd size: The buffaloes' position with respect to wind direction and the presence of woodland. As the buffaloes spread around the rim of the floodplain, their various groups become differently situated with respect to wind direction, and those elements of behavior that are affected by olfactory perceptions are subject to more variable stimuli than when most of the buffaloes were centrally placed in an open area and were uniformly affected by the wind. The wind may then become an indirectly disruptive agency for group behavior, rather than an indirectly unifying one.

The second factor is density of woodland. The tendency of *Syncerus* is toward forming smaller groups when in woodland, regardless of population density. This is well substantiated in West Africa. Unfortunately, West African buffaloes occur in population densities that are nowhere sufficiently large to provide exhaustive proof. Sinclair (1977), whose field experience is mainly in East Africa, also states, as a fact, that herds are smaller in forested habitats in spite of a high density of individuals. In the case of *S. c. caffer* in open Central or East African woodland, this does not mean the formation of very small groups but it does mean that very large groupings, that is, composite herds such as still occur in some open grasslands, are no longer present. In a wooded habitat, if herds of several hundred are present, relatively minor events, such as a single attack by lion which causes part of the herd to flee while the rest remains stationary, or the finding of a water-hole that is too small to accommodate all the individuals, may lead to a split which will result in two herds. These two situations have been observed more than once by me in Zambia and Zimbabwe. In wood-land habitats, the response by portions of herds to the leadership init-iative of individuals appears more pronounced than in open grass-lands, which also contributes to the reduction of herd sizes.

At the end of the rains, as flooded grasslands begin to dry out, the herds tend to return to the flats. This commences before the ground ceases to be soggy and flooded locally. During the transition from wet to dry I have had to follow buffalo herds for many kilometers through water 15 to 60 centimeters in depth. The buffaloes never stopped for any length of time in areas that were even slightly submerged. They could not experience any shortage of food in these semiflooded locali-ties, and the reason for avoiding flooded areas appears to be an aver-sion by buffaloes to resting and ruminating in water. Possibly an acces-sory factor may be the hindrance to calves of water that is more than a few centimeters deep.

Three types of herd regimens have been discussed above. They are linked to (1) an open grassland habitat with water available throughout; (2) a well-watered woodland habitat; and (3) any situation where regular commuting between pasture and water is necessary. The first type al-lows for the largest herds, while the other two lead to smaller herds. The third type necessitates the greatest degree of herd discipline.

Age and sex composition of herds

Herds of *S. c. caffer* normally contain fewer adult males than adult females. The ratio of adult males to adult females can be as high as 0.9:1, the two sexes almost approaching numerical equality, or as low as 0.4:1, the number of males less than half that of females. These differences in the adult male-to-female ratios can be partly explained by the separation of some bachelor males from the herds. Additionally, both native and foreign hunters tend to kill more males than females. Other reasons for the numerical superiority of adult females are con-ceivable, for example, sex-linked susceptiblity of males to diseases. The mortality rate resulting from combat between buffalo males is so low as to be probably negligible in this context. Sinclair (1977) argues convinc-

Table 4.4. *Average percentages of various age categories in herds of* S. c. caffer *from several localities*

Age category	Lolgorien, Kenya (1965)	Mabalauta, Zimbabwe (1972)	Busanga, Zambia (1974)	Kruger National Park, South Africa (Pienaar, 1969)[a]
Infants	17.3	⎱17.5⎰	13.0	8.2
Juveniles	7.9		8.6	6.8
Subadults	17.3	15.8	11.4	13.8
Adult males	22.4	24.6	19.6	32.7
Adult females	35.4	42.2	47.2	38.5

[a] Average weighted for the numbers of individuals involved, of the Shabin, Makamba, and Bob herds, portion of Sabi-Sand herd, and a random sample of the Crocodile Bridge population, all from Kruger Park.

ingly for the probability of an equal sex ratio at birth in the African buffalo, using his own data and those of others.

A buffalo population may usefully be divided into the following age categories:

infants	0–1 year old
juveniles	1–2 years old
subadults	2–4 years old
young adults	4–5 years old
vigorous adults	5–about 14 years old
old adults	about 14 years old or older

The division between vigorous adults and old adults probably cannot be defined simply in terms of time, as climatic, alimentary and possibly parasitological factors may influence the effective rate of aging. The six age categories proposed above can be reduced to four by combining the three adult divisions.

Table 4.4 shows the proportions of age categories in herds, irrespective of sex except in the case of adults. The figures are average percentages for several localities in which I have made counts and include data calculated from Pienaar (1969a, p. 45) for the Kruger Park.

Groupings and associations within herds

The greater part of a buffalo herd consists of basic herds, which may be in contact with one another or mutually separated by distinct breaks. There may be considerable interpenetration along the edges of adjacent basic herds, so the concept of "mobile territory" cannot be strictly applied to the area occupied by a basic herd at a given time. However, a tendency to stay apart has been observed. Interpenetration is seldom complete because the "intruders" are reluctant to stray far from their own center, whereas the members of the "invaded" basic herd have a tendency to filter inward or laterally toward their own greatest concentrations. Strays are not usually forcefully expelled from basic herds to which they do not belong, although this does occur now and then.

The gravitation of individuals toward their basic herds, within the context of a large herd, is perhaps best observed at times of danger, when the herd becomes alert and flees. Whenever this happened with herds in which I was able to identify a large number of individuals, an unmingling of basic herds was indicated during the early phase of a stampede, irrespective of whether they continued by running in separate bodies or squeezed together, as they may tend to do when pursued by lions. There were inevitably a few strays, which hurriedly sought out their basic herds immediately after a stampede, partly guided by vocalization. It may be appropriate to explain that buffaloes often run only for short distances, of a few hundred meters, and it was occasionally possible to follow them sufficiently closely to observe the termination of a stampede.

Next most prominent in the herd are the bachelor male clubs. Up to 55 percent of the adult males live within these groups, which may either accompany the herd or break away from it for long periods. This chapter, however, is directly concerned only with those bachelor groups that remain with the herd. It is common for them to be located on the flanks of herds, in clusters or long narrow formations. There is frequently a bachelor group near the front of a herd and one behind it, sometimes lagging as much as 300 or 400 meters in the rear. Because of this peripheral distribution of bachelor clubs it is easy to get an exaggerated impression of the number of males in a buffalo herd, as they tend to mask the bulk of the females from a ground observer. Those bachelor clubs that remain attached to a herd may be larger than isolated bachelor clubs (see the earlier section, "Group size") and have a more unstable membership than the latter. The proportions of subadults in bachelor clubs that stay with herds are normally higher than in bachelor clubs which have separated. In the Lolgorien herd, between 37.5 and 55.5 percent of the subadult males lived in bachelor clubs. In the Busanga herds, about 40 to 60 percent did so. Although isolated bachelor clubs only rarely include a female, those that stay with herds are commonly infiltrated by a few females. This infiltration, however, results in very few social interactions.

Besides these groupings, a herd may comprise others of a less essential nature. I have observed in Kenya (Lolgorien), Zimbabwe (Mabalauta), and Zambia (Luangwa and Zambezi valleys) that when juveniles are numerous, they may form groups led by a female, sometimes accompanied by one or two other females keeping in the background. The leaders of five juvenile clubs observed in detail were:

a vigorous cow with a big calf of her own – 1 case;
a mature vigorous cow with strong secondary male traits – 1 case;
an old cow – 2 cases;
an old cow with strong secondary male traits – 1 case.

Five to sixteen individuals were observed to compose these groups, including the leader, any other adult females, and sometimes one or more young subadults. Such juvenile clubs were usually located somewhere toward the middle of a herd but have been observed to stray forward, especially on resumption of herd grazing after a rest period.

At the rear of herds, especially large ones, groups may form composed of sick, old, and disabled individuals, sometimes females with very young calves, and lost calves. Social behavior was seldom observed within such invalid groups. They were one of the two types of observed buffalo groups that were aggregates rather than social entities.

The other groups that lacked internal social behavior were small aggregates of individuals that were active in guiding herds when they were on the move, which I call *pathfinders*. These were loose aggregates, consisting occasionally of as many as a dozen individuals of either sex and variable age.

Thus at its most complex a buffalo herd may contain a large number of component groups. For example, the Buffalo Bend herd in the Mabalauta area, Zimbabwe, when numbering about 400, consisted of 11 groups of 15 to 60 individuals, including a juvenile club, three bachelor male clubs, and 7 mixed groups (basic herds).

In a large buffalo herd the different group boundaries may become diffuse. However, the groups, particularly basic herds and bachelor clubs, cannot be said to disintegrate even when they become indiscernible, because of the rapidity of their reappearance and the constancy of their memberships. Unless we accept them as permanent features of the buffalo social structure we must imagine that the herds exist at a kind of equilibrium, on the one side of which component groups vanish, to reappear on the other in all their complexity, like a reversible chemical reaction. This does not seem to be acceptable, given the observed speed with which distinct groups reappear.

In general, the African buffalo herd pattern is largely the consequence of the presence of a strong tendency to affiliate in groups, without a correspondingly strong tendency to forcibly expel intruders.

5

Herd movements

Migration

Syncerus caffer does not make regular seasonal migrations, such as are made by the East African wildebeest (*Connochaetes taurinus albojubatus*) and several other African ungulate species. However, indications exist in the folklore and in a few actual observations that the buffalo did migrate occasionally, and perhaps may do so even now, given the right incentives and opportunities. For example, in 1926 a northward migration occurred from the Northern Rhodesian (now Zambia) district of Kasempa (Pitman, 1934). The reason for this migration cannot now be ascertained. Prior to the rinderpest epizootic of the 1890s, buffaloes and other Bovidae that shared the same habitat were much more numerous, and grazing pressures on favorable land were necessarily much greater than in later years. There were no human installations to impede long-distance trekking in search of fresh pasture, and it may be that buffaloes were more migratory than they are now. Thus the springbuck (*Antidorcas marsupialis*), formerly known for mass migrations (e.g., Shortridge, 1934), ceased to migrate in most of its range, owing to its reduced numbers and the appearance of man-made obstructions. Even if buffaloes did make occasional long treks in the past, however, it seems doubtful that these were regular seasonal events.

Home range

At present, isolated buffalo herds are normally confined to home ranges that they use continuously for years. In regions where several herds coexist, the exact boundaries of their home ranges may shift slightly with the years, just as the herds themselves may recombine to some extent, as observed in the Busanga and Chunga areas, Zambia,

63

1971–78, the southwestern end of the Loita region, Kenya, 1965–67, and in the Gona-Re-Zhou area, Zimbabwe, 1972.

The size of a buffalo herd's home range varies according to the abundance of edible grass and available water, as well as other more subtle factors such as physiographic features and both intraspecific and interspecific competition. Among my various size estimates of buffalo home ranges, nine considered to be the most accurate vary between 126 and 1,075 square kilometers, while buffalo population densities within home ranges vary between 0.17 and 3.77 per square kilometer. The nine estimates represent three regions: one in Kenya, one in Zambia, and one in Zimbabwe. Home ranges of different herds in the same region are not necessarily of similar size, and correlations of herd and home range size are weak. Where buffalo populations are sufficiently large to result in the formation of more than one herd, home ranges may sometimes overlap. In the Busanga flats of Zambia the home ranges of six herds were traced in the dry season of 1974 (Figure 5.1). They occupied a total of 4,625 square kilometers, and overlaps covered about 9.5 percent of the area. There were three separate areas of overlap and herds were observed to meet and mingle in two of these areas. In the Mabalauta area, Zimbabwe, in October 1972 an overlap of between 10 and 20 percent existed in the home ranges of three herds, two of which are known to have mingled during that time. A small overlap was also shown by Sinclair (1977) to occur in two Serengeti herds, although he stresses the smallness of this overlap and does not describe any mingling between herds.

Herd movements within the home range

A herd establishes a route within its home range and follows it over and over again. Since the main reason for this movement is to procure sufficient graze, the route is a strip of country whose width varies, according to the size of the herd, from under 1 kilometer to as much as 5 kilometers. Within this strip, though occasional deviations may occur, many places are visited each time the herd passes by. These places include watering spots as well as exceptionally favorable grazing and resting localities. They also include licks associated with saline concentrations in termite mound material and salts in muds near some springs. Also visited are localities in which physiographic features and vegetation offer particularly good protection by their effect on visibility, airborne scents, and sounds, probably in that order of importance, although the relative importance of sight and smell may be sometimes reversed.

The degree of adherence to the general route partly depends on the density of edible grasses within the home range. If the area consists mainly of poor pasture, as for example the Mabalauta sector of the

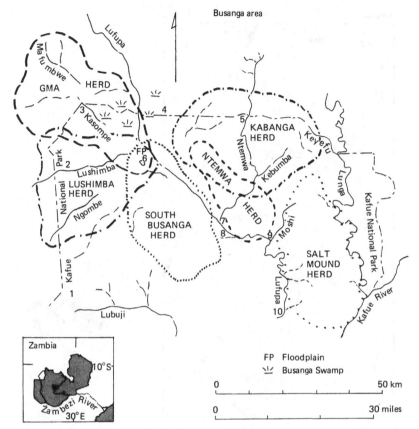

Figure 5.1. The July 1974 ranges of the main buffalo herds in the Busanga area, northern boundary of the Kafue National Park, Zambia, as determined by detailed tracking and direct observation. The range limits are known to have slightly altered at various times during the 1970s, bringing the entire population into mutual contact at some time, as range overlaps changed. Individual herd compositions are also known to have undergone occasional changes. The Ntemwa herd was the most stable. The ranges of the Kabanga and Salt Mound herds often overlapped. The extensive floodplain of the Lufupa and Lushimba rivers around the Kasolo Kampinga thicket (6 on the map) is usually an area of herd overlap, where the majority of herd meetings, compositing, and exchanges were observed. The area is covered by open woodland, open floodplain grassland, grassland/*dambo*, and swamp. All of these, usually with the exception of true swamp, are utilized by buffalo herds. Key: (1) Chitokotoko Game Guard; (2) Lushimba G.G.; (3) Kasompe G.G.; (4) Masozhi G.G.; (5) Kabanga, G.G.; (6) Kasolo Kampinga thicket; (7) Ntemwa camp; (8) Big Tree camp; (9) Moshi camps, old and new; (10) Lufupa camp.

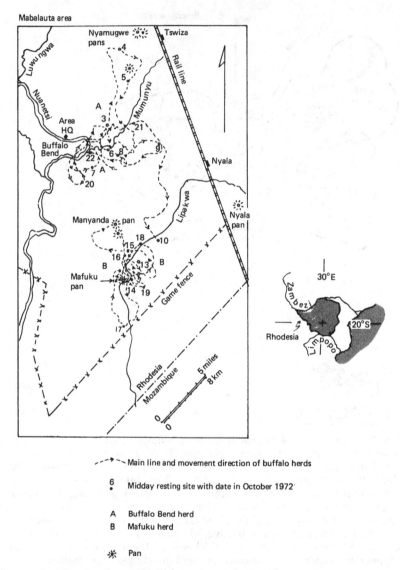

Figure 5.2. The October 1972 range of the Buffalo Bend and Mafuku herds, Mabalauta area, Gona-Re-Zhou Game Reserve, southeastern Zimbabwe. The area is a low-hill and valley scrubland, with best grazing near streambeds and very localized and reduced quantities of drinking water in the dry season.

Gona-Re-Zhou Game Reserve in Zimbabwe, the search for graze at the end of the dry season causes so many deviations along the route that it is disrupted, except in its broadest outline. Nonetheless, for that particular herd in October 1972 (Figure 5.2) an examination of some hardened old tracks from the previous wet period indicated that with improved grass conditions the route became more constant, as well as longer. An illustration of this was observed when it rained during the night of October 3–4, 1972 and the herd reacted immediately by extending its route into an area which, by the evidence of old tracks as well as reports, was included in the main circuit, but where water had dried out. The herd moved 9.5 kilometers overnight to reach the area, but excluded it again from its route as soon as the water from the isolated rainfall dried out. A herd's degree of adherence to its established route also depends on the amount and timing of predation by man and lion, in that order of importance. Regular and moderate predation do not cause a herd to deviate greatly from its route, but heavy and irregular predation, particularly by man, brings about a very apparent increase in stress among the buffaloes that often leads to erratic movements and even to herd disintegration, as observed in the southeastern corner of Kenya between Kilibasi, Kuraze, and Lungalunga, 1966–68, parts of Kenya's Narok, Samburu, and Laikipia districts, 1964–67, and several localities in western Zambia, from the Zambezi Valley in the south to the Lukwakwa and Mayou *dambo* locations in the north, including the edges of the Kafue National Park.

A herd that is not under stress from an abnormally high level of disturbance, once it is committed to reach a certain place, for example, a watering or grazing locality, does not alter this commitment except in cases of extreme interference. This was observed in Lolgorien, Kenya, 1964–65, Kafue National Park, Zambia, 1970–78, the Zambezi Valley and Gona-Re-Zhou, Zimbabwe, 1971–72. For example, when a moving herd is harried by lions, it may double back along its path or deviate sharply from its preattack direction of movement, and often run intermittently for several kilometers. Once out of danger, the herd stops for a short time, until the healthy stragglers catch up with it and a normal herd shape is restored, and then movement is resumed. The herd moves at some large angle to its flight direction, usually at a rapid pace, for up to 3 or 4 kilometres, changing direction often. It breaks into a run occasionally, apparently when some alarming scent reaches it, but it is not possible to ascertain from my observations whether these occasional runs are exclusively a response to lion scent, a response to any unfriendly or unknown scent, or simply because of nervousness. The threshold of fright is evidently low at such a time. At the end of a few minutes to several hours of this behavior, the herd faces toward its prealarm destination, even though the latter may have to be approached from a direction substantially different from that taken in normal circumstances. Eventually, the herd rejoins its usual route. In July 1974 this behavior was a major object of my study, in the northern sector of the Kafue National Park. Paths followed by buffalo

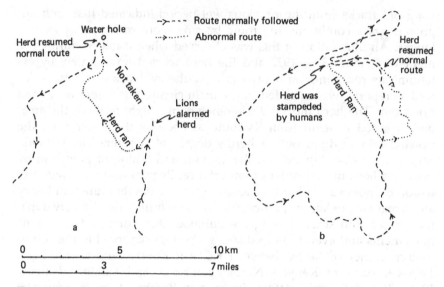

Figure 5.3. Portions of actual routes followed by buffaloes, illustrative of the persistence with which buffalo herds will adhere to habitual patterns, even when interfered with. (a) Lions interfered with a moving herd; (b) a herd was intentionally stampeded by humans. In both cases, after taking evasive action, the buffaloes returned to their habitual paths. The terrain in both cases was rather flat woodland, grassland, and *dambo* in the Kafue National Park, Zambia, immediately east of the Busanga area. (Dry season, 1974.)

herds after lion attacks were traced and a herd was caused to stampede from a position along its established route, and its subsequent path traced (Figure 5.3a,b). In all cases the herds were back on their habitual routes within 8 hours at the most.

Routes established by buffalo herds are circuits which may be simple imperfect ovals or complex shapes with crossovers and loops. They are sometimes not circuits but reciprocating movements along an axis. The habitual resting and watering places often lie slightly off the main route and upwind from it under prevailing wind conditions. The observed lengths of buffalo herd circuits vary between 50 and 105 kilometers, but these lengths should not be thought of as limiting cases.

Wind direction and herd movement

The relationship of a herd's direction of movement and the wind direction is variable, despite an observed tendency to walk into the wind, either head-on or obliquely. In 36 observations on apparently displaced and disoriented individual buffaloes or groups, 31 instances (86%) of initial movement were upwind. In the remaining 5 instances, 1 featured wind blowing downslope and 2 featured material barriers in the upwind direction; the last 2 cases cannot be qualified in any way. In buffaloes that have not been temporarily disoriented, by finding themselves in anomalous circumstances, a tendency to move upwind may also exist, but if it does it is much weaker. There is no general synchronicity between the directions of wind and buffalo movements. In most of *S. c. caffer*'s range the air currents tend to have the prevailing easterly component of the trade winds, and only in the season of the passing of the intertropical front (i.e., during rainy seasons) is there a likelihood of variable winds. The directions of purely local draughts, which may be more variable, may affect the direction of a herd's movement on a very small scale but are not simply correlatable with the overall movement patterns. The displaced buffalo's tendency to walk into the wind makes it appear probable that the circuit or axis retraced by a buffalo herd is initially established by following the scent of water and grass on the wind. But once established, the circuit or axis of movement itself appears to become the prime control, regardless of wind direction. Thus terrain or landmark recognition appears to play an important role. A buffalo herd forced by lions to deviate from its habitual route finds its way back to it, even if this entails downwind movement. In any area and season, it is virtually certain that either a circuit or an axis of movement has segments that must be negotiated downwind.

Rates of movement

Except during resting periods, a buffalo herd is mostly in linear or arcuate motion through its home range, at rates varying from about 0.15 kilometer per hour, when grazing intensively, to about 9 kilometers per hour when walking rapidly. Of 83 measured rates of herd movement, in Kenya, Zambia, and Zimbabwe, which probably approximate a random sample, 65 percent fell between less than 0.1 and 0.6 kilometer per hour, which gives an average of 0.3 kilometer per hour (Figure 5.4). Assuming 18 hours of movement daily, this gives a range of 1.8 to 10.8 kilometers per day, or an average of 5.4 kilometers per day. This agrees well with the generally estimated rates of movement, by less precise means. The fast walk is uncommon in the buffalo and forces it into a special posture with the withers low, neck outstretched, and nose close to the ground or straight out.

Buffalo herds never run except when fleeing from an immediate threat or, exceptionally, to water. A running herd with calves has been clocked in open grassland at 44 kilometers per hour for a distance of 1.1 kilometers, after which all the animals continued walking at a rapid pace (Zambia 1977). Running speeds of 36 to 42 kilometers per hour were recorded on several occasions in buffalo herds. Alexander et al. (1977) noted that the maximum speed of the East African buffalo is considerably less than the maximum speed of antelopes. Elsewhere, a maximum speed of 56 kilometers per hour, was recorded (in Schaller, 1972, p. 233), but I would not expect it to be attained frequently.

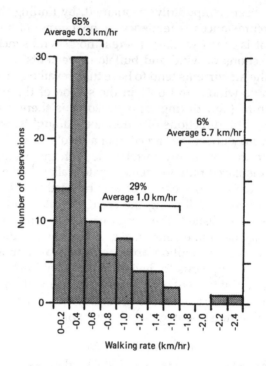

Figure 5.4. Rates of herd movement, measured in Kenya, Zambia, and Zimbabwe. In addition to the rates shown above, there were 1 observation at 7 km/hr, 1 observation at 8 km/hr, and 1 observation at 9 km/hr. Fast walking is uncommon in the African buffalo.

Occasionally, the rate of herd movement across the home range may drop to almost nil for a number of days. If there is a particularly favorable locality on a herd's circuit, or if grass fires affect the locality toward which the herd is moving, it may spend several days circling within a small area and watering at one place.

Configurations in moving herds

Buffalo herds move in columns and in wide-front formations. Both patterns are common in herds of all sizes. At times, in very large herds and in composite herds, the buffaloes may be formed into more than one column, advancing in some kind of nonrepetitive *en echelon* pattern. In open country, the column formation is adopted when relocation takes clear precedence over grazing. In woodland, herds often retain a column formation even when feeding takes precedence over forward movement, presumably because of the interference of the trees. In such cases the spacing between individuals is increased mani-

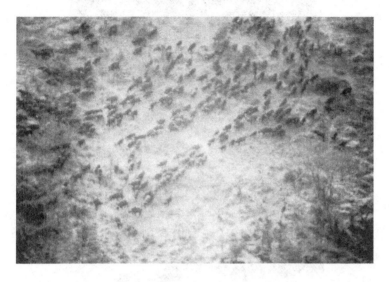

a

b

Figure 5.5. (a) A herd photographed from the air, showing internal groupings, probably basic herds. (Southern Tanzania.) (b) A large herd on the march in column (the front of the Busanga composite herd).

fold, out of grazing necessity, but if alarmed or if signals to move, for example, to water, are vocalized, the buffaloes compact quickly into a walking column. Columns are 1 to some 10 animals wide, sometimes organized in files, and there are often breaks between basic herds or other components of the herd (Figure 5.6a). There are pathfinders in front, usually with a gap between them and the other buffaloes. Figures 5.5b and 5.6a,b illustrate the column formation. (See also Figure 5.8a.)

a

b

c

Figures 5.6. (a) A herd on the march in column, showing the kind of break that often occurs between basic herds. (Western Zambia). (b) A herd walking in column formation, with four pathfinders in front and the top male last in the small advanced cluster. (c) The same herd as in (b) passing the same place on another circuit (note small isolated bush on both photographs), but walking in a wide-front formation, with a pathfinder on its left wing and slightly in advance. The right wing of the herd does not appear on the photograph. (Western Zambia.)

When a large herd moves in a loose column through woodland or bush, contact among the members decreases, and collective actions, such as changing direction, lose the benefit of the buffalo's innate herding instinct operative in compact formations or where visibility is good. In woodland, the turning of the herd as a whole is brought about by several individuals in conjunction with the pathfinder in the lead. This behavior was followed on four occasions in Zambia, in the Busanga and Ntemwa herds during 1974 and in the Chunga and Ntemwa herds during 1977, and observed casually in others. As the herd moves forward, either purposefully or grazing at a low rate, the leading pathfinder walks in a small arc until it faces the new direction and halts for a moment, looking ahead in the alert or scanning posture (see Figure 8.3a,b). The animals behind are thereby slowed down to a near-halt. Almost immediately, a nearby animal on the side of the buffalo column opposite to the turn direction, that is, on the outside of the future curve in the route, swings in the same manner as the path-finder and walks across the column to its opposite edge, at an ordinary pace, always followed by several animals, which in turn causes others nearby to similarly change direction. Soon after, another animal on the far side of the herd turns, crosses over, is followed, and this is contin-ued along the column with other animals initiating the turning move-ment among their near neighbors. At the end of this activity, all these assistant pathfinders are on the opposite side of the column from that on which they started, and the entire herd now faces in the new direction, although throughout it the buffaloes are never all in sight of one another (Figure 5.7). Since most of the individuals that assisted in making the turn had been seen on other occasions to perform pathfind-ing functions, the activity appears to be a "rehearsed" way of turning a herd when visibility is obstructed, with the least amount of disruption. It may occur even when the pathfinder turns while not in front but elsewhere in the herd, in which case the portion of the herd preceding the pathfinder may move on straight ahead for a short distance and then gradually loop back and rejoin. By the time all the buffaloes have turned, the leading pathfinder has moved, usually grazing, fairly far ahead, inducing momentum in the new direction. It was evident from examinations of old spoor that turns were not made at the same spots or at the same angles, even along well-established routes. One obser-vation was made in a locality that was not on the herd's habitual route. Therefore, a hypothesis that this behavior is a habitual collective re-sponse to known landmarks is untenable, and a decision-making ele-ment appears to be present.

The wide-front formation is often seen in open country and is mainly adopted when relocation does not have clear precedence over grazing. It is not simply a dispersal of individuals in order to obtain grazing

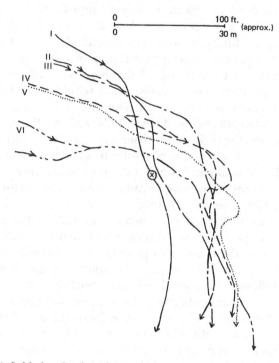

Figure 5.7. A field sketch of tracks made by a turning herd. This typical turning movement was initiated by a single buffalo whose advance is shown by line I. This animal halted in an alert stance at the point marked **X**, then walked on as indicated, followed by a string of other buffaloes. The initiative was taken up by other individuals, each followed by a string of buffaloes. The numbering of the tracks represents a time sequence, over a period of about 10 min, by the end of which the entire herd had turned. The herd, numbering some 300, was walking through open woodland. Chunga area, Kafue National Park, Zambia.

space but an ordered walking formation adapted to grazing requirements. This formation is only a few animals deep but many animals in width: A herd of 400 may have a front of some 30 or more individuals. It is compact when walking, with the appearance of an unbroken mass, but when forward movement slows down the spacing between individuals almost immediately increases, both along and across the direction of forward movement, and the herd is ready to graze as it advances at a very slow rate. Pathfinders are located in front or on a flank, usually upwind. The wide-front formation is illustrated in Figure 5.6c.

When buffaloes, moving in a wide front, cross from open country into woodland and keep on walking, they as a rule re-form into a column. This is one of the situations in which basic herds may stand out, as they separate in order to follow one another, instead of walking more or less abreast. In very many cases, however, herds enter wood-land immediately prior to settling down for a rest, and do not pene-trate deeply but halt near the edge. In such cases, there is a loosening up of the wide-front formation but columns do not form. At certain times, especially when young foliage appears on woodland vegetation as the first showers of an approaching rainy season descend, herds may enter woodlands to browse. In such cases, browsing begins at the edge of woodland and the wide-front formation becomes looser as the browsing buffaloes penetrate the treed area. If a herd is alarmed in this situation, it finds it very difficult to reorganize, which may lead to confused congested gatherings, disjointed stampedes, and occasionally the splitting of a herd.

Distribution in moving herds

When a herd moves in column, it is practically always preceded by one or more pathfinders, with a gap between them and the other buffaloes. All or some of the high-status males may come next, and near them are a few individuals of either sex which at other times function as path-finders and remain alert most of the time. Basic herds follow, either separated by gaps or overlapping, and with bachelor males strung out along the column's flanks. They are followed by invalids, which may be spread over a considerable distance, especially if the herd is walking rapidly, and bachelor males are mingled with them, singly or in groups. Those high-status males that are not in front may be trailing the herd. Females that are seen sometimes scanning the country be-hind the herd are not permanently in the rear but belong to some basic herd. Usually they are individuals that also function as pathfinders (see Figure 5.8b).

In a wide-front formation, the animals in the forward line are a mixture of high-status males, high-status females from the basic herds, more dominant bachelor males, and occasionally other ani-mals, followed in fair but imperfect order by members of their own respective hierarchies. One or more pathfinders may walk ahead of the herd's front, as in a column, but this is not always so and very frequently the leading pathfinders are located slightly forward on the upwind limb of the herd. From above, the wide-front formation usu-ally appears crescentic and concave on the advancing edge, with a "tail" at the rear. A most spectacular modification occurs when the high-status bulls precede the front of the herd in a closely packed parallel line (see Figure 5.8c).

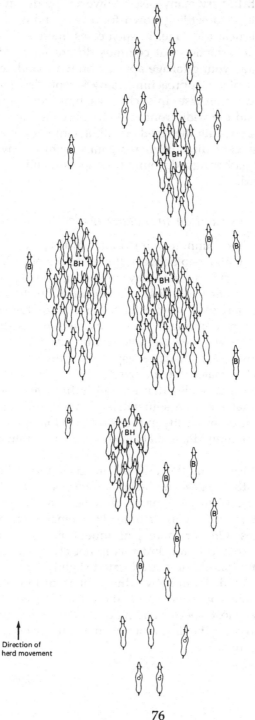

Direction of
herd movement

a

76

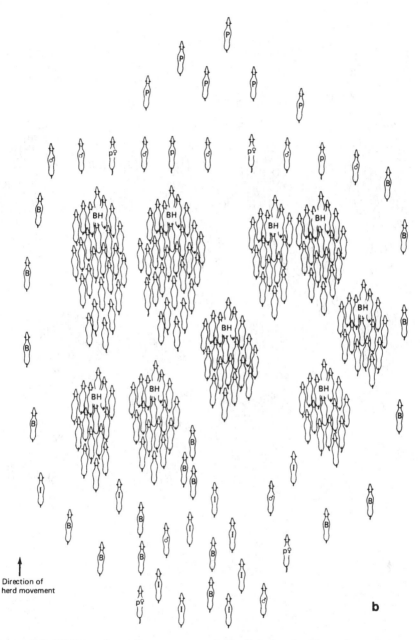

Direction of
herd movement

b

Figure 5.8. Walking formations common in buffalo herds: The column formation (a) is commonly adopted when relocation takes clear precedence to feeding, and in wooded country. The wide-front formation (b and c) is often seen
in herds moving across open ground and when relocation does not take clear
precedence over grazing along the route. Key: P, pathfinders; p, assistant
pathfinders; p♀, assistant female pathfinders; ♂, high-status males; B, bachelor
males; ♀, high-status females; I, invalids and aged; BH, basic herd.

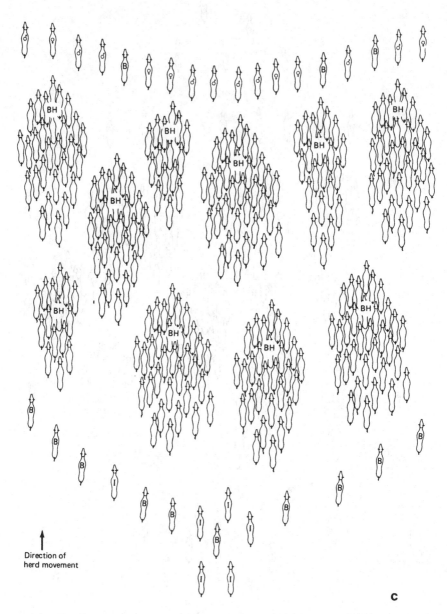

Direction of
herd movement

c

Figure 5.8 (cont.)

6

Feeding activities and alimentary processes

Grazing

Throughout most of the range of *S. c. caffer*, the seasonal changes in pastures have characteristics in common. The main dry season is the time when widespread pastures become depleted and grass eaters search for green remnants near water or in depressions, or make do with dry low-nutrient grasses. Grass fires are common in the dry season and are followed by an almost immediate regrowth of perennials, which at that stage become a preferred, though sparse, pasture. During the early rains, annual grasses increase in quantity and are heavily grazed (See Figure 6.1a.). As the rainy season advances, perennials grow tall and have relatively high nutrient content in their upper parts. At the height of the rainy season, these grasses are both prolific and nutritious, although they are not all equally acceptable to the buffalo or to other species (Figures 6.1b and 6.2). At the end of the rains, but before drought sets in, nutrients in the stems and leaves begin to migrate to the roots of the perennials, to remain stored there during drought and reducing the food value of the vegetative parts of these grasses (Figure 6.3). Seed heads may retain seeds and provide some high-nutrient food for part of the dry season.

Mitchell's (1963) compilation of grasses and sedges of the Kafue National Park and adjacent areas supplies some feeding records. These records, supplemented slightly during the present study, include 56 grass and sedge species (plus unidentified *Bothriochloa* and *Hyparrhenia* spp.) on which mammals have been observed to feed. Buffaloes were recorded feeding on 31 of these (plus unidentified *Bothriochloa*, *Brachiaria*, *Hyparrhenia*, and *Jardinea* spp.), as shown in Table 6.1, with some seasonal preferences. Table 6.2 lists the mammals observed to feed on the same grass species as *S. c. caffer*. In the Kafue National Park,

a

b

Figure 6.1. A herd grazing on (a) the first grass of the wet season and (b) the luxuriant pasture in advanced wet season. (Zambia.)

Figure 6.2. A grazing female in a relaxed posture, chewing. Note grass protruding from the mouth. (Wet season pasture.)

Figure 6.3. Two subadults grazing on sparse dry season pasture. Note the extent of tail movement during tail switching. Two oxpeckers (*Buphagus erythorhynchus*) sit on the nearer buffalo's back.

Table 6.1. *Grasses and sedges reported eaten by S. c. caffer*

Genus	Tsavo, Kenya[a]	Tarangire, Tanzania[b]	Serengeti, Tanzania[c]	Manyara, Tanzania[d]	Mount Meru, Tanzania[e]	Ruwenzori, Uganda[f]
Andropogon						
Aristida						
Botrochloa						spp.
Brachiaria						
Cenchrus		ciliaris				
Chloridion						
Chloris						gayana
Cynodon		dactylon	dactylon	dactylon	dactylon	dactylon
Cyperus				laevigatus	laevigatus	
Dactyloctenium						
Digitaria	macroblephora	macroblephora				
Echinochloa						
Eleusine						
Enneapogon						
Eregrostis	spp.	spp.				
Heteropogon						
Hyparrhenia						filipendula
Jardinea						
Loudetia						
Oryza						
Panicum	coloratum, maximum		coloratum, max- imum, infestum			
Pennisetum			mezianum		landestinum	
Phragmites						
Pogonarthia						
Rhynchelytrum						
Schizachyrium						
Schmidtia						
Setaria			chevalieri			
Sorghum						
Sporobolus			spicatus, pyr- amidalis	spicatus, pyr- amidalis		pyramidalis
Themeda			triandra		triandra	
Trachypogron						
Typha						
Urochloa						
Vossia						

[a] Leuthold, 1972, cited in Sinclair, 1977, p. 305.
[b] Lamprey, 1963, cited in Sinclair, 1977, p. 305.
[c] Sinclair, 1977.
[d] Vesey-Fitzgerald, 1969, cited in Sinclair, 1977, p. 305.
[e] Sinclair, 1977.
[f] Field, 1968a, b, cited in Sinclair, 1977, p. 305.
[g] Grimsdell, 1969.
[h] Bourlière and Verschuren, 1960.
[i] Vesey-Fitzgerald, 1960. The question mark (?) in entries in this column indicates my inference that *S. c. caffer* eats the given species but a direct statement to this effect is lacking in the text. Several other grasses are similarly listed by

(G.)	Queen Elizabeth, Uganda	Congo (Zaire)[h]	Lake Rukwa, Tanzania[i]	Kafue Zambia[j]	Zambezi Valley, Zimbabwe[k]	Wankie, Zimbabwe[l]	Mabalauta, Zimbabwe[m]	Kruger, South Africa[n]
(An.)				*gayanus, amplectens*		*gayanus*		*gayanus, amplectens*
(Ar.)						*pilgeri*		
(Bo.)		spp.		spp.				spp.
(Br.)				*brizantha,* spp.				*nigropedata*
(Ce.)				*ciliaris*			*ciliaris*	*ciliaris*
(Chn.)				*cameronii*			*cameronii*	
(Chs.)			*gayana?*	*gayana*				
(Cyn.)			*dactylon?*	*dactylon*		*dactylon*		
(Cyp.)			*usitatus*			*laevigatus*		spp.
(Da.)					*gigantum, sp.*		*sp.*	
(Di.)				*milanjiana, gazensis*			spp.	spp.
(Ec.)			*pyramidalis*	*colonum, stagnina, pyramidalis*		*?*spp.		*stagnina*
(El.)				*indica*				
(En.)						*cenchroides*		
(Er.)				*superba*	*rigidior rigidor*	*pallens, superba*	*rigidior,*	*superba*
(He.)				*contortus*			*contortus*	*contortus*
(Hy.)		*filipendula,* spp.		*filipendula, dissoluta, rufa,* spp.	*filipendula, dissoluta*	spp.	*filipendula, dissoluta*	*dissoluta*
(Ja.)				*sp.*				
(Lo.)				*superba*		*flavida*		
(Or.)			*barthii?*	*barthii*				
(Pa.)			*maximum?*	*maximum*			*maximum*	*coloratum, maxiumum*
(Pe.)								
(Ph.)				*mauritianus*				*communis*
(Po.)						*fleckii*		
(Rh.)				*setifolium*				
(Scz.)						*sanguineum*	*sanguineum*	
(Shm.)							*bulbosa*	*bulbosa*
(Se.)				*pallide-fusca, porphyrantha, sphacelata*	spp.			*flabellata*
(So.)								*verticilliflorum*
(Sp.)	*pyramidalis*	*pyramidalis*	*spicatus, robustus, marginatus?*	*pyramidalis*				*robustus*
(Th.)		*triandra*		*triandra*				
(Tr.)				*spicatus*				
(Ty.)			*sp.?*					*capensis*
(Ur.)				*bolbodes*				
(Vo.)			*cuspidata*	*cuspidata*				

Vesey-Fitzgerald: *Cyperus edulis, Diplachne fusca, Paspalidium geminatum, Scirpus maritimus,* and *D. jaegeri.* Some of these were probably eaten by *S. c. caffer* but no positive statement appears. Since Vesey-Fitzgerald comments that some grasses were avoided by the 18 herbivorous species he lists near Lake Rukwa, I did not include these grasses in Table 6.1.

[j] Mitchell, 1963; present study.
[k] Present study.
[l] Wilson, 1975.
[m] Present study.
[n] Pienaar, 1969.

Table 6.2. *Numbers of grass/sedge species recorded as eaten by herbivores of the Kafue National Park, Zambia*

Mammal species	Number of grass/ sedge species shared with *S. c. caffer*	Total grass/ sedge species eaten
Waterbuck, *Kobus defassa*	17	22
Puku, *K. vardoni*	14	16
Wildebeest, *Connochaetes taurinus*	13	25
Warthog, *Phacochoerus aethiopicus*	13	19
Hartebeest, *Alcelaphus lichtensteini*	11	15
Burchell's zebra, *Equus burchelli*	10	16
Hippopotamus, *Hippopotamus amphibius*	9	15
Impala, *Aepyceros melampus*	9	13
Reedbuck, *Redunca arundinum*	8	13
Oribi, *Ourebia ourebi*	4	6
Lechwe, *Kobus leche*	4	5
Sable, *Hippotragus niger*	4	7
Roan, *H. equinus*	3	7
Elephant, *Loxodonta africana*	3	4
Spring hare, *Pedetes capensis*	2	4
Bushbuck, *Tragelaphus scriptus*	1	2
Eland, *Taurotragus oryx*	1	3
Cane rat, *Thryonomys swinderianus*	1	1

Source: Mitchell, (1963); present study.

waterbuck (*Kobus defassa*) is the most similar to buffalo in the grass species consumed. This was also true around Lolgorien, Kenya during the 1960s. However, despite a moderately strong preference for the same grasses, buffaloes and waterbucks were seldom observed grazing simultaneously in the same location. The most important food grasses of *S. c. caffer* in the Kafue region of Zambia appear to be:

Throughout the year	*Echinochloa stagnina*
	Hyparrhenia filipendula
September–April (dry and hot to wet)	*Echinochloa pyramidalis*
	Trachypogon spicatus
December–April (wet)	*Andropogon gayanus*
	Brachiaria brizantha
	Chloridion cameronii
	Digitaria milanjiana
	Echinochloa colonum
May–August (cold)	*Vossia cuspidata*
September–November (hot)	*Hyparrhenia dissoluta*

The most generally important of these grasses appear to be *E. stagnina*, *E. pyramidalis*, *D. milanjiana*, *T. spicatus*, *H. filipendula*, and *H.*

Table 6.3. *Grass zones in the Lake Rukwa area, Tanzania, showing species
also present in the Kafue National Park, Zambia*

Zone (Lake Rukwa)	Grass species (Lake Rukwa)[a]	Present in Kafue N.P.[b]
Perimeter of plains	*Hyparrhenia rufa*	X
	Chloris gayana	X
	Sporobolus marginatus	
Depressions, watercourses	*Vossia cuspidata*	X
	Oryza barthii	X
Alkaline lakeshore grasslands	*Sporobolus robustus* dominant	
	Diplachne fusca or	X
	Cyperus laevigatus dominant	
	Sporobolus spicatus and	
	Diplachne jaegerii	
Lakeshore deltas	*Cynodon dactylon*	X
	Paspalidium geminatum	
	Typha sp. (during floods)	
Woodlands	*Panicum maximum*	X
	Digitaria sp.	?
Acacia parkland	*Cyperus usitatus*	
	Sporobolus marginatus	
	Echinochloa colonum	X

[a] After Vesey-Fitzgerald (1960).
[b] After Mitchell (1963).

dissoluta, not only because of their high acceptability to many grass
eaters but also because of their available biomass or seasonality.

Near Lake Rukwa in southwestern Tanzania, a flood–drought area
which in several ways resembles Zambia's Busanga flats and periphery, the grasses that Vesey-Fitzgerald (1960) particularly identified with
respect to the buffalo are *Vossia cuspidata,* from the end of the rains
(April) onward through the dry season, and *Echinochloa pyramidalis.*
The early rains in November lure the buffalo into acacia parkland to
feed on *Cyperus usitatus* and unspecified annuals. Vesey-Fitzgerald
listed zonally the grasses that occur in the Rukwa valley and referred to
them as "pasture." Undoubtedly many of these grasses were eaten by
buffaloes. However, he listed 18 herbivorous mammalian species for
the Rukwa area and treated their utilization of pasture largely as a
whole. He also stated that some tall perennial grasses appeared to be
"sour," that is, avoided when mature or dry. The grass zones in the
Lake Rukwa area indicating those species of grasses also present in the
Kafue National Park, Zambia are listed in Table 6.3.

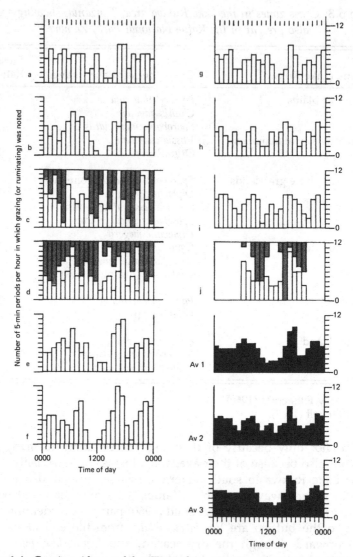

Figure 6.4. Grazing (denoted by □) and rumination (denoted by ▓) in individual buffaloes. Observation interval: 5 min. Histograms a to j, each for a separate individual, show the number of times grazing; ruminating times in three cases (C, D, J) were noted during each hour. A to F were observed in the dry season and G to J in the wet season. All were adults. A, B, C, E, and I were males, and D, F, G, H, and J were females. Their localities are given in the text. Grazing averages (■): Av1 = dry season, Av2 = wet season, Av3 = overall average.

The 18 herbivores listed by Vesey-Fitzgerald as feeding on (some of) the grasses listed for Lake Rukwa in Table 6.3 are as follows (where * indicates species also present in the Kafue National Park):

Elephant, *Loxodonta africana**

Buffalo, *Syncerus caffer caffer**

Hippopotamus, *Hippotamus amphibius**

Puku, *Kobus vardoni**

Topi, *Damaliscus korrigum*

Burchell's zebra, *Equus burchelli**

Bohor reedbuck, *Redunca redunca*

Eland, *Taurotragus oryx**

Giraffe, *Giraffa camelopardalis*

Impala, *Aepyceros melampus**

Warthog, *Phacochoerus aethiopicus**

Roan, *Hippotragus equinus**

Hartebeest, *Alcelaphus lichtensteini**

Waterbuck, *Kobus defassa**

Bush duiker, *Sylvicapra grimmia**

Bushbuck, *Tragelaphus scriptus**

Steinbok, *Raphicerus campestris**

Aardvark, *Orycteropus after**

The grasses known to be commonly eaten by buffaloes in other parts of East Africa are listed by Sinclair (1977, p. 305) and here included in Table 6.1. The following among them occur not only in these locations but also in the Rukwa and Kafue areas:

> *Chloris gayana* (wet regions)
> *Cynodon dactylon* (ubiquitous)
> *Panicum maximum* (dry savanna)

Sporobolus pyramidalis is tabulated by Sinclair as common buffalo graze in East Africa, and Grimsdell (1969) notes that this is a preferred species in Uganda.

Grasses observed eaten by buffaloes in the Zambezi Valley and the Wankie and Mabalauta areas, Zimbabwe, as well as Pienaar's (1969a) list of grass species preferred by the buffaloes of Kruger National Park appear also in Table 6.1. *Panicum maximum* is reported as part of the buffalo's diet from both Kruger, South Africa, and Mabalauta, Zimbabwe, which together with the other reports documents it as a species very widely used by *S. c. caffer*. *Themeda triandra* is another widespread species, widely grazed by the buffalo, both in the south and toward the north of its present range. It has been shown of importance in the Serengeti (Sinclair, 1977) and west of the great lakes (Bourlière and Verschuren, 1960), as a food source for buffaloes.

Observations of single animals

I observed nine adult individuals, each for uninterrupted periods exceeding 24 hours, during which the presence or absence of grazing activity was recorded at 5-minute intervals (about the shortest intervals that I could sustain recording over long periods). The nine individuals were:

male, Kenya, Marallal area,   dry season, 1966 (Figures 6.4a, 6.5a)

female, Kenya, Loita area,   wet season, 1967 (Figures 6.4g, 6.5g)

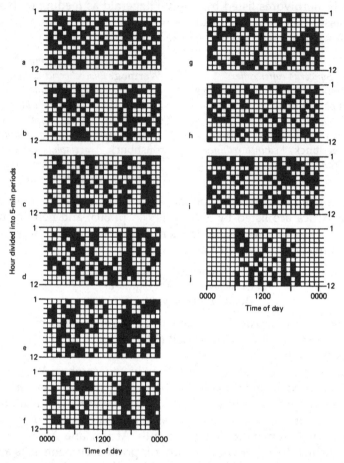

Figure 6.5. Grazing in individual buffaloes. The data presented in Figure 6.4 are here replotted to show the distribution of grazing activity within each hour. Black squares indicate 5-min periods in which some grazing was noted.

male, Zimbabwe, Mabalauta area,	dry season, 1972	(Figures 6.4e, 6.5e)
female, Zimbabwe, Mabalauta area,	dry season, 1972	(Figures 6.4f, 6.5f)
male, Tanzania, Ruaha area,	dry season, 1973	(Figures 6.4c, 6.5c)
male, Zambia, Busanga area,	dry season, 1974	(Figures 6.4b, 6.5b)
female, Zambia, Busanga area,	dry season, 1977	(Figures 6.4d, 6.5d)
male, Zambia, Chunga area,	wet season, 1977	(Figures 6.4i, 6.5i)
female, Zambia, Chunga area,	wet season, 1977	(Figures 6.4h, 6.5h)

Also one female in the Busanga area was observed for a 15-hour period, in the end of the wet season of 1973 (Figures 6.4j, 6.5j). These animals include three for which complete ruminating records were obtained during the same observation periods. In plots of the distribution of grazing behavior observed in these individuals over the 24-hour cycle (Figures 6.4 and 6.5) similarities are apparent. Recording of grazing behavior for individual animals during long unbroken periods was found easier than recording rumination.

Direct observation of individual grazing rates is made difficult by the height of wild grasses, which hides a grazing buffalo's muzzle. It has to be done at close range. At very close range the sound of a buffalo's bite on grass can easily be picked up on a tape recorder. In most cases, about 40 bites per minute were counted, as compared to 50 to 70 bites per minute for domestic cattle.

Observations of groups

Records of the presence or absence of grazing behavior were extracted from notes on various buffalo groups from western Zambia. Taken together, they represented all 24 hours of the diurnal cycle. Because the numbers of records available for each of the 24 hours were unequal, excess records above the minimum number available were discarded, on grounds of reliability and conditions under which the observations were made. Notes that were retained represent observations with an average interval of about 20 minutes. The compilation of these data resulted in a set of 18 sequences, each covering the 24-hour cycle, 13 of them representing the dry and 5 the wet season. I made the assumption that the chances of noting any grazing behavior that occurred during each of these 432 hours (i.e., 18 × 24 hours) were equal and therefore a plot of the observed grazing frequencies (Figure 6.6) would be a valid approximation of the actual grazing pattern. Figure 6.6 shows similarities to Figure 6.4, which illustrates the daily grazing for animals observed individually.

The nine 24-hour records of single animals were grouped into a dry season set and a wet season set and plotted in this form (Figure 6.7a,b). The grazing was plotted as a percentage of time for each half-hour, to parallel Sinclair's (1977, p. 84) figure 24 for Serengeti buffaloes.

Daily duration and distribution of grazing

Grazing occupied between 8 hours 35 minutes and 10 hours 20 minutes of the day (Figure 6.4a–j), that is, 36 to 43 percent of the buffalo's time. It is the single most time-consuming activity of the buffalo. The daily grazing total tends to be slightly less during the wet season (Figure 6.4).

The most important grazing time in the dry season is between 1500 and 1800 hours. During this 3-hour period the observed occurrence of grazing is not only high but shows a progressive statistical increase

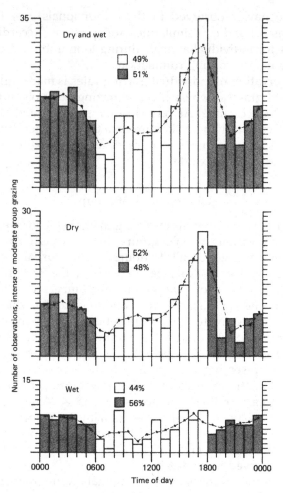

Figure 6.6. Grazing in groups. Each hour of the diurnal cycle was observed on an equal number of separate occasions, and the observation interval averaged about 20 min (was never less than 15 min). Key: □, full daylight; ▨, night and transition periods.) The broken line is a three-item moving average, which is introduced to cover minor fluctuations and accentuate a similar location in time of grazing highs and lows during both dry and wet seasons. However, the late afternoon grazing high is much greater in the dry than in the wet season. During the latter, grazing is in general more evenly distributed. The first and last points on the moving-average graph were obtained by assuming a 24-hr repetition of the histogram. The data are from western Zambia.

until about 1730 hours. Toward the end of this period, a sharp drop in grazing activity takes place. Grazing starts to increase again, though not abruptly, at some time between 1900 and 2000 hours and proceeds at frequencies that average around 47 percent of the total time between

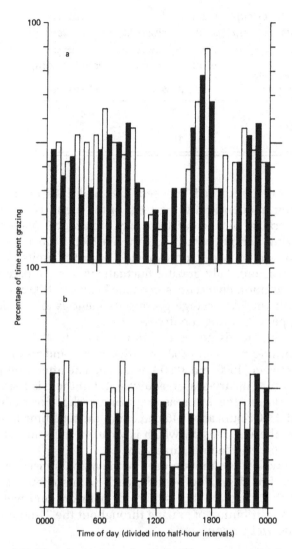

Figure 6.7. The proportion of each half-hour spent grazing by adult buffaloes of both sexes, based on an average of six individuals observed for 24 hr during the dry season (a) and three individuals observed for 24 hr during the wet season (b). The animals are the same ones as those represented in Figures 6.4 and 6.5, A to F and G to I. The observation interval was 5 min.

2000 and 0330 hours. At some time between 0330 and 0600 hours, a drop in grazing occurs, most often shortly before sunrise. Grazing again increases afterward, often during the dawn period, and tends to reach a high level some time between 0800 and 1000 hours. The observed frequency of grazing during this last interval, although not

Table 6.4. *Group grazing rates in S. c. caffer, expressed as percentages of all observed grazing activity*

Observed rate of group grazing	Percentage of total observed grazing	
	Dry season	Wet season
Intense	58	43
Moderate	24	24
Low	18	33

nearly as great as that recorded between 1500 and 1800 hours, and matched by grazing frequencies during parts of the night, is nevertheless noteworthy because of its consistent daily occurrence. Aside from the preferred grazing periods, which lie between 1500–1800 and 0800–1000 hours and which coincide approximately with the limits of daylight, no really consistent daily grazing fluctuations were apparent. A major resting period normally occurs some time between 0900 and 1500 hours, which results in low average grazing frequencies and which is the major resting period during the dry weather.

Wet-season grazing tends to occur with a consistently high frequency in the morning between 0800 and 0900 hours and toward the end of the day between 1500 and 1800 hours. The late-afternoon peak is not nearly as high as its dry-season equivalent, but there is a similarity between the dry and the wet season grazing cycles. For example, between 2000 and 0330 hours about 46 percent of the total time is spent on grazing during the wet season, which is virtually the same as during the dry season (see preceding paragraph).

Three grazing lows – at around sunrise, midday, and sunset – generally separate three broad grazing periods, each with sometimes pronounced internal fluctuations. This is true for both dry and wet seasons, although the distribution of grazing throughout the 24-hour cycle is more even in the wet season.

Grazing rates (group)

The rate of group grazing was classified in three "present" categories and one "absent" (nil) category (see Table 6.4). The classification is not rigid and leaves much to the observer's judgment, especially since intermediate conditions are possible. However, it provided a useful framework and I seldom felt uncertainty about classifying grazing activity within these categories. A truly quantitative assessment of group grazing rates is impossible in the field without the aid of short-interval oblique photography.

Table 6.5. *Grazing during full daylight hours compared with grazing during night and transition hours observed in S. c.* caffer

		Percentage total grazing	
Sample	Season	0600–1800	1800–0600
6 Adults, both sexes, average	Dry	49	51
3 Adults, both sexes, average	Wet	49	51
Group grazing, average	Dry	52	48
Group grazing, average	Wet	44	56
Average of the above samples	Both	49	51

Intense grazing was recorded when all or nearly all group members grazed, paying little attention to anything else. *Moderate* grazing was recorded when about half of a group grazed most of the time. *Low* grazing was recorded when substantially less than half of a group grazed at any one instant, often erratically, with an inclination to be easily distracted from grazing or to stop altogether. Grazing was recorded as *nil* when the statement was literally true, but occasional very brief token grazing by one or two individuals was ignored for statistical purposes.

The seasonal changes in grazing rates shown in Table 6.4 are in harmony with the more even distribution of grazing frequencies in the wet season.

Day and night grazing

I have no clear evidence that night grazing is more common in *S. c. caffer* than day grazing. My figures show a not-entirely-consistent trend toward more grazing during the 12 hours from 1800 to 0600, but the difference is very small (see Table 6.5).

Sinclair (1977, p. 82) states, "Although African buffalo graze mainly at night, they graze during daylight hours as well" while he also says (1977, p. 90) that "a significant increase in daylight grazing can be seen during the dry season (p = .002, Mann–Whitney test)." His figure 29 (1977, p. 89) gives, in its bottom third, curves for grazing by day and night during wet and dry seasons that suggest subequal emphasis on grazing by day and night over the year as a whole. However, these curves show a marked change from mainly nighttime grazing in April and May to mainly daytime grazing in June to August. These data disagree with the much smaller differences between night and day grazing indicated by my own observations in the various localities listed at the beginning of this chapter. Sinclair's curves are based on only three pairs of comparable plots for day and night grazing (March–April, July–August, and August–September) and plots for grazing by night in December and by day in February–March that are unpaired. No plots or graphs are given for October or November. Thus the only comparable data cover the period March–April to August–September.

It is my impression that usually the amounts of day and night grazing are approximately equal, but occasional divergencies may favor either time depending on special local conditions, of which the most common are hunting and interference by man, which inhibit daylight grazing.

Effects of weather on grazing

The effects of weather on grazing patterns of *S. c. caffer* and domestic cattle (*B. taurus*) are similar. In his studies of the grazing behavior in domestic cattle, Hancock (1953) notes that intensive grazing activity occurs during breaks between storms or rainshowers. Conversely, grazing is stopped at times of heavy rain or strong wind, as well as at the peak of a storm. *S. c. caffer* herds have been observed to act in a similar way.

The greatest differences in the condition of individuals belonging to the same herd have been observed in regions with long dry seasons, toward the end of the latter. At those times, some adults may appear to have maintained good condition, while the skeletal parts of others may be clearly showing under their skins, apparently as a result of malnutrition. Thus Hafez and Bouissou's statement, supported by other sources, regarding domestic cattle, that "Inherent individuality is an important factor in grazing" (Hafez, 1975, p. 208), appears to hold also for the African buffalo. I noted in the Mabalauta area, Zimbabwe, where dry-season pasture is generally poor, very marked differences in grazing selectivity among individuals of a herd. Whereas some moved steadily forward making only a cursory selection and taking most of the grass within reach, others circled slowly over a small area, often several times, carefully selecting only a fraction of the available grass. The latter individuals generally showed poorer condition.

Browsing

Browse has been reported to constitute between 6 and 55 percent of *S. c. caffer's* total food intake. The former figure is from Tarangire, Tanzania (Lamprey, 1963), and the latter from the Zambezi valley (Jarman, 1971). My own field estimates, excluding observations in parts of the

Zambezi valley, range from some 3 to some 14 percent of the total intake. The estimates are based on the time spent browsing by selected individuals, the proportion of individuals browsing during a given period of time, the amount taken per bite while browsing as compared to grazing, and 16 on-the-spot stomach content examinations, including individuals whose feeding patterns were known. The lower of these estimates is from southwestern Kenya when grass was plentiful, and the higher is from southeastern Zimbabwe at the end of the dry season when grazing was poor.

Buffaloes browse for one of two reasons: (1) Selective feeding or (2) grass shortage. In Lolgorien and Entasekera, Kenya, it was normal under good pasture conditions for large numbers of buffaloes to browse simultaneously just before dark, that is, some 2 hours after their main resting period. In the Kafue region of western Zambia, such concentrated browsing was not observed regularly when good graze was plentiful, but many individuals of both sexes and all ages above 6 months spent some time browsing independently, mainly on and near termite mounds, during grazing periods or at the end of resting. These types of browsing cannot be explained by a lack of pasture, and have never been estimated to exceed, and seldom attained, 9 percent of the total intake. They must, therefore, represent selective feeding, for the purpose of obtaining specific nutrients, possibly metals, or a particular form of roughage, and perhaps both.

Under poor pasture conditions, such as for example occur at the end of the dry season in parts of Zimbabwe's Gona-Re-Zhou Game Reserve (which includes the Mabalauta area), buffaloes may browse in bulk when opportunity arises, eating their way from one end of a shrub area to the other as they advance, in the same manner as they do when they graze intensively. In extreme cases, such as may occur in the Zambezi valley where grass-poor mopane woodlands and *Commiphora–Combretum* thickets cover much of the area, browse may become a condition of survival, and Jarman's estimate must be of the right order. In general, however, such high levels of browsing are abnormal for *S. c. caffer*.

My observations suggest that browsing is a variable, dependent on pasture conditions, increasing as the latter deteriorate, and thus ultimately on the buffalo's ability to select its range, that is, on man-induced land restrictions. At the same time, there is a strong indication that a small amount of browse, of the order of 5 percent of the total food intake, is sought by buffaloes even under the best pasture conditions. Table 6.6 shows plant species other than grasses definitely recorded as eaten by *S. c. caffer*, by the sources quoted, including the present study.

Table 6.6. Shrubs, trees, and herbs reported eaten by S. c. caffer

Species	Location							
	Serengeti Tanzania[a]	Ruwenzori, Uganda[b]	Albert N.P., Congo (Zaire)[c]	Kafue N.P., Zambia[d]	Zambezi Valley, Zimbabwe[e]	Wankie N.P., Zimbabwe[f]	Mabalauta, Zimbabwe[g]	Kruger N.P., South Africa[h]
Acacia sieberiana				X[i]				
Afzelia quanzensis				X	X			
Annona stenophylla				X				
Aspilia kotschyi				X				
Baphia massaiensis				X		X		
B. obovata						X		
Bauhinia macrantha				X				
Burkea africana				X				
Capparis tomentosa		X						
Colophospermum mopane				X	X		X	X
Combretum ?celastroides					X			
C. hereroense								
C. spp								
Commiphora spp					X	X	X	X
Dichrostachys cinerea					X	X	X	
Dicliptera nemorum				X				
Diospyros mespiliformis				X?				
Diplorhynchus condylocarpon				X				X
Euclea divinorum				X				X
Euphorbia oatesii				X				
Grewia bicolor	X						X	X
G. messinica								X
G. monticola								X
Heeria insignis								X
H. reticulata				X				

Species						
Justicia betonica		X				
Karissa edulis		X				
Lonchocarpus capassa		X				
Melanthera scandens		X				
Parinari capensis		X[i]				
Paropsia brazzeana		X				
Plectranthus cylindraceus		X				
Pluchea ovalis			X			
Pterocarpus antunesii				X		
Sclerocarya caffra					X	
Securinega virosa						X
Sesbania sesban	X	X				
Strychnos cocculoides		X				
Terminalia mollis		X				

[a] Sinclair, 1977. [b] Field, 1968b. [c] Bourlière and Verschuren, 1960. [d] Mitchell, 1963; present study. [e] Present study.
[f] Wilson, 1975. [g] Present study. [h] Pienaar, 1969. [i] Fruit also eaten.

Drinking

Drinking in the dry season

Observations during the dry season indicated that 42 percent of all daily drinking occurred between 1600 and 1900 hours. This 3-hour period constitutes the main peak in drinking activity observed during the dry season (Figure 6.8). Eleven percent of all drinking took place between 0900 and 1100 hours. This constitutes a small peak in drinking activity since the expected proportion for any 2-hour period is 8.3 percent. Sixty-one percent of all dry-season drinking was observed to occur during the full daylight period, between 0600 and 1800 hours.

The African buffalo of the *caffer* subspecies drinks at least once a day, with very few known exceptions. Consequently, during the dry season, the main framework for a buffalo herd's movements is provided by water locations. Buffaloes tend to drink between dawn and late morning and again between about 1600 hours and dusk, but pasture conditions and protective behavior may exert a strong influence on drinking times. Surface water may become scarce with the advance of the dry season, causing many grazers to roam in the neighborhood of the few remaining sources of drinking water, depleting the nearby pastures, which suffer particularly heavily if hippopotamus is abundant. This depletion of waterside pastures may induce herds of large animals needing much graze, such as buffaloes, to wander far from water. Dry-season grazing away from water usually requires roaming over larger pasture areas than wet-season grazing. Thus the necessity to walk two, three, or more times as far as under better grazing conditions may cause a buffalo herd to arrive late at the watering place, while a substantial daily variability in the distance between water and the herd's last pasture may lead to considerable differences in the time of arrival. In localities where a serious threat of human predation exists either at the watering site or along the way to it, buffaloes may delay coming to water until dark. These factors may also cause other changes in a herd's drinking behavior, for example, from drinking twice to drinking once a day.

Some grazing always intervenes between resting and drinking. Even if after resting a herd moves in a straight line toward water, the movement begins slowly, accompanied by moderate or intense grazing. The grazing drops off as the herd nears the watering site, a characteristic vocalization may then be sounded, and the herd walks the rest of the way to the water without grazing. If a buffalo herd rests by a watering site, it normally moves from the water, grazing, at the end of the rest, then returns to drink. The grazing bout between resting and drinking varies widely in duration, occasionally lasting only a few minutes.

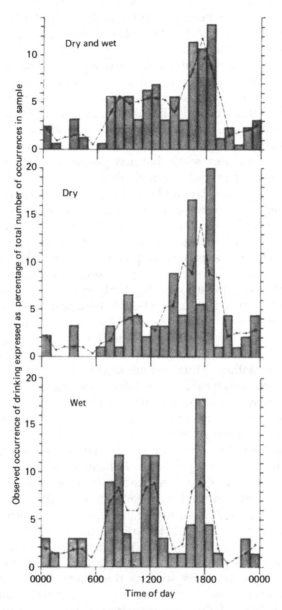

Figure 6.8. Drinking throughout the diurnal cycle. The three-item moving-average curve (broken line) emphasizes the greater amount of drinking during daylight hours in both seasons. A morning and an early evening drinking activity peak occurs in both seasons, but whereas in the dry season there is an unequaled increase of drinking in the early evening, the activity is more evenly distributed throughout the day in the wet season. There is equal-observation time distribution within each graph. Dry season n = 90, wet season n = 67.

After drinking, a buffalo herd usually walks at least a short distance away from the water before it begins to graze. This behavior may be due to poor grazing around the edges of the water reservoir, where grass tends to be trampled, but it may also be a response to other factors, which build up stress and discourage a prolonged stay. Thus if a herd drinks from a reservoir which is too small to accommodate all the animals at once – a common occurrence – the forward animals, pressed from behind by those that follow, are induced to drink fast and leave the water immediately afterward. Furthermore, other species tend to converge on drinking sites at about the same times of day, and interspecific encounters result that may put stress on the buffaloes. For example, I have repeatedly seen elephants interfering with buffaloes' drinking, by threat displays and direct aggression. Black rhinos (*Diceros bicornis*) sometimes do the same, although their present small numbers make these interactions uncommon, with possibly only the exception of the Luangwa Valley reserves in eastern Zambia. The neighborhood of drinking sites is frequented by predators, and I have observed lion attacks on buffaloes in such places on several occasions. Crocodiles may present a real threat to wading buffaloes in some larger bodies of water, and in Zambia I have seen buffalo groups drinking from shallow muddy waterholes situated near cleaner, easily approachable but crocodile-infested low-velocity streams. I could only explain it as a measure to avoid crocodiles. In some cases watering sites are a focus for human predation. Thus various sources of stress, which may be expected to promote tension in buffaloes, are associated with watering sites. Stress may begin to build up even before the drinking place is reached, for if water and pasture are widely separated, as often happens in the dry season, then a large stress-promoting gap is to be expected between the start of a need state and its satisfaction, with respect to drinking. The presence of stress from this source is made evident when buffalo groups run to water, a behavior that I observed in Kenya's coastal hinterland (1967) and southeastern Zimbabwe (1972). The speed with which buffalo herds often complete watering and the frequently observed tense or alert behavior in individuals, even when no other species are in evidence, suggest that buffaloes in fact often feel under stress near watering sites, and therefore may be expected to leave them as soon as they have drunk. On the other hand, in localities where few threats are associated with watering sites, particularly if the buffaloes are very familiar with the area, herds may go so far as to start the main resting period close to the water, almost immediately after the morning drink. The buffaloes then often remain in that locality, to drink from the same reservoir in the afternoon. Dry-season watering behavior may be therefore modified by the buffaloes' familiarity with any particular water reservoir. However, with a few exceptions, drinking appears associated with more stress than most grazing. As soon as the buffaloes put a little distance between them and the watering site after drinking, tension appears to dissipate and cognitive behavior is much less in evidence.

Groups of 178 to 850 (average = 464) individuals were observed in the dry season to take from 14 to 77 minutes to drink, with an average of 43.3 minutes ($n = 40$). The average time per individual derived from the above, that is, 0.09 minute, is not useful in itself because it contains variables. Theoretically, group drinking should be amenable to quantitative treatment. In any particular case, the total herd drinking time (T) could be expressed in an equation incorporating the average drinking time per individual (t), the average drinking area required per individual (a), the number of individuals (N), and the total available drinking area (A). Then, where A is less than Na,

$$T = \frac{tNa}{A}$$

and where A exceeds Na,

$$T = t$$

In practice this relationship could be applied to predict approximately the length of time a group will take to drink only where those features that uniquely modify water reservoirs, for example, accessibility from the edges and range of depths, could be so assessed as to give a realistic value of A, and the number of drinking periods per day could be established.

The observed length of time required for an adult individual to drink its fill when watering once a day ranged from 4.4 to 9.6 minutes ($n = 21$) with an average of 6.3 minutes. Pienaar (1969a) gives rough estimates of water intake by individual adults from the Kruger National Park, South Africa, based on direct weighings of stomachs with contents, from animals that were darted on their way to drink and others immediately after having drunk. He gives averages for two group estimates, 31.3 and 30.46 liters. He also states (1969a, p. 39) that "In another case 40.0 Kg water was obtained from the wet stomach of a cow that had just drunk her fill." He clearly states that one of these cases is a daily estimate, and therefore all three are here inferred to be daily intakes. The average of these three figures is 34 liters. In conjunction with my average for observed individual drinking times (6.3 minutes), it is possible to estimate the average adult water intake at 5.4 liters per minute (= 90 milliliters per second). This rate is some 2 to 2½ times that possible for an adult human. The daily water intake of Zebu cattle (*Bos indicus*) in California's Imperial Valley was reported by Ittner et al. (1951) as 10 U.S. gallons, that is, 37.9 liters, which is similar to *S*.

c. caffer, as compared to the Hereford breed of *Bos taurus*, which consume 16 U.S. gallons, that is 60.6 liters, daily in the same environment.

In several cases buffalo herds were observed to approach water at a good walking pace, barely slowing down when they waded into it, drink very quickly, and come out again, usually on the far side if it was a waterhole or a shallow stream. The drinking, which is in general mainly a sucking and not a lapping operation in *S. caffer*, was then done loudly with the head tilted slightly more upward on the lowered neck than when drinking at ease. It is possible that the amounts of water drunk in this manner are comparable to those drunk normally, despite drinking times that are obviously shorter than those quoted in the preceding paragraph. Since the volumes drunk in this rapid way are unknown, drinking rate estimates cannot be attempted. I was unable to establish a clear correlation between this special drinking behavior and any other events that could have brought it about. However, all the buffalo groups that occasionally drank in this way were among those that I considered in general to be particularly prone to tenseness or flight, probably because of the observed high frequency of threat state in their habitat.

Buffaloes that water more than once a day may drink for only 45 seconds to about 2 minutes at one time, as has been observed in a Mabalauta (Zimbabwe) herd in 1972. In these cases a pattern that is frequent throughout the area of my observations consists of drinking at dusk or a little earlier, grazing within a kilometer or so of the water reservoir, and returning to drink again between 2030 and 2300 hours.

Drinking in the wet season

During the wet season 79 percent of all drinking was estimated from observations to occur during the period from 0600 to 1800 hours. The period from 1200 to 1800 hours accounts for 58 percent of all drinking, with two peaks, one of which occurs between 1200 and 1400 hours, representing 24 percent of all drinking, and the other between 1700 and 1800 hours, representing 18 percent of all drinking. Another preferred drinking period occurs from 0700 to 1000 hours, and represents 20 percent of all drinking.

If water is present more or less throughout the pasture areas, in pools and streams, herds of over 100 often do not drink as single bodies but instead tend to do it in groups, at slightly different times and perhaps at different, though not widely separated, watering sites, of which there may be several in one locality. Thus in a large herd, grazing, resting, ruminating, and drinking may take place simultaneously in the wet season. Uniform behavior is less important when pasture and water occur together, since the need to travel from one to the other does not then exist. In many cases, however, even in the wet season good drinking water is localized, and then buffalo herds usually drink in a body. Herds of 112 to about 900 (average = 440) individuals were observed in the wet season to take from 10 to 72 minutes to drink, with an average of 35 minutes ($n = 16$). These figures resemble their dry-season equivalents.

Drinking intervals

In any case, a basic 24-hour cycle, characterized by a watering bout approximately at dusk, but with much leeway, appears to exist. Water intake within this basic cycle, however, may be divided into several drinking periods, with less water drunk during any one of them. The spacing between these drinking bouts within the 24-hour basic cycle is somewhat irregular, with the exception of a rather persistent and consistent drinking peak in the morning. Thus a 24-hour drinking cycle may be the intrinsic interval, prone to modifications by extrinsic factors.

Anomalous drinking behavior

Exceptions to the 24-hour, or shorter, drinking interval appear to exist. Vesey-Fitzgerald (1960) writes about the Rukwa area in southwestern Tanzania: "Although buffalo will regularly visit water holes when available, the herds can remain for an indefinite period on the green pastures which become available in the dry season due to regrowth of *Echinochloa* after fire, or on the trampled *Vossia* meadows, *without drinking*. During the hot days of October and November, however, they *may* seek shade at noon" (emphasis added). Vesey-Fitzgerald's statement is made certainly with full awareness of the buffalo's normal drinking habits, as clearly indicated by the division of the Rukwa herbivores into 'shaders" (reedbuck, topi, eland) and "drinkers" (zebra, buffalo, puku, hippo, elephant), an important distinction in that area, where shade and surface water do not coexist in the dry season. In the Ruaha National Park, situated some 300 kilometers to the east, in the dry season of 1973, I saw buffaloes at least 23 kilometers from the Great Ruaha River, north of the escarpment, with no sign of nearby surface water.

Licking

Six positive observations were recorded of adult buffaloes (four males and two females) licking the stems of tall thick grasses, such as *Echinochloa pyramidalis* or *Pennisetum purpreum*, in the upward direction, when in semidry condition.

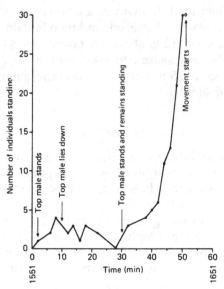

Figure 6.9. End-of-rest behavior. The plot shows the transition from complete rest to forward movement in a buffalo herd. A moment after the 30th individual stood up, the entire remainder of the herd, which was lying down until then, stirred and general movement began. (GMA herd, Busanga, Zambia, November 1977.) Compare with Figure 6.10.

Buffaloes lick salt (mainly NaCl) incrustations formed around some springs. They also lick termite mounds apparently for the salt content, particularly the smaller, grey, relatively indurated types, a few of which, located on habitual buffalo routes, may be conspicuously eroded by this activity. I have occasionally observed buffaloes lick one another's bodies when covered in mud from wallowing.

Resting

The main body of data on resting behavior was gathered in Zambia and Zimbabwe. In those countries the year is fairly sharply divided into one wet and one dry season. For wild ungulates, the most important single seasonal factor is the effect that rainfall has on pastures, so my seasonal limits are drawn between humid and dry pasture conditions, rather than at the start and finish of annual precipitation. Humid pastures occur in Zambia and Zimbabwe from November to about the end of May, although the transitions from wet to dry, and vice versa, vary somewhat from year to year and from locality to locality. In the data presented below, May and November are included with the wet season.

A buffalo herd may rest at any time of day or night but not with equal frequency at all times. The favorite resting time throughout the

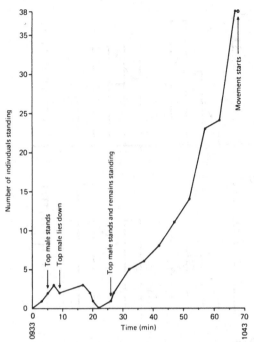

Figure 6.10. End-of-rest behavior. The plot shows the transition from complete rest to forward movement in a bachelor bull club of 38 individuals (Busanga, Zambia, October 1971). Compare with the end-of-rest behavior in a buffalo herd shown in Figure 6.9.

year is between 1200 and 1600 hours, while the least amount of resting occurs during the transitions from night to day and day to night. The main difference between the seasons is found in the amount of contrast between the frequencies of resting behavior at different times. The ending of a major resting period is signaled by a high-status animal, usually the top male, by standing up, starting token movement and grazing, and sometimes vocalizing (Figures 6.9 and 6.10; see also Figures 6.11–6.13).

The greater number of observations was made in daytime. For this reason and also because of the necessarily inferior reliability of night observations, the data were compiled in two separate sets, one for the full-daylight period from 0600 to 1800 hours, and another for the night and transitional periods, from 1800 to 0600 hours.

Resting in the dry season

The duration of resting periods in the dry season ranged from 12 minutes to 5 hours 55 minutes, with an average of 1 hour 50 minutes.

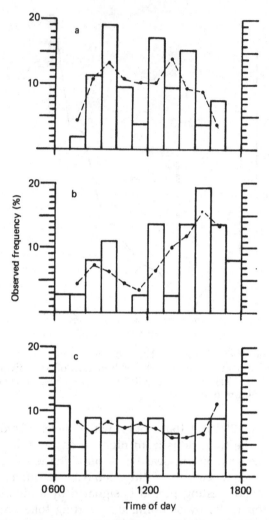

Figure 6.11. Observed diurnal distribution in Zambian and Zimbabwean buffalo herds: (a) start of resting; (b) start of movement at the end of resting; (c) walking accompanied by little or no grazing. The broken lines are three-item moving averages of the histograms.

Normally there was one major resting period during the day, very often another major one during the night, and several shorter resting periods scattered throughout the 24-hour cycle, at intervals varying between half an hour and 3 hours.

Of the total time the buffaloes spent resting, about 53 percent was during daylight, here defined as between 0600 and 1800 hours. They seldom rested in the early morning but very often in the early after-

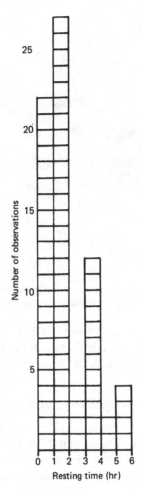

Figure 6.12. Observed length of resting in buffalo herds from Kenya to Zimbabwe.

noon: 42 percent of observed daylight resting (22 percent of all resting) was in the 3-hour span from 1200 to 1500 hours, about 1.7 times the statistically expected amount. On the other hand, between 0600 and 0900 hours, only 5 percent of the observed daylight (3 percent of all) resting occurred, which is about one-fifth that which may be statistically expected.

Forty-six percent of the resting observed during the hours of darkness and transition, from 1800 to 0600 (22 percent of all resting), was in the 4-hour period from 0100 to 0500 hours, and 13 percent (6 percent of all resting) from 2200 to 2300 hours. Little resting behavior was ob-

Figure 6.13. A herd moving at the end of the midday rest. In the foreground an individual is entering a muddy waterhole to wallow. (Zambia.)

served from 0500 to 0600 and from 1800 to 1900 hours, altogether 8 percent of the observed night and transition (4 percent of all) resting, divided about equally between the two periods (see Figure 6.14a,b).

Resting in the wet season

The observed duration of resting periods ranged from 45 minutes to 3 hours 32 minutes, with an average of 2 hours 6 minutes. The tendency for a major rest to take place in the afternoon was less strong than in the dry season. Other resting periods were more nearly equal in length than in the dry season.

Of the total observed resting time, 51 percent occupied the full-daylight part of the diurnal cycle. This is more nearly the statistically expected value of 50 percent than the dry season value, but it seems unlikely that the slightly greater dry season divergence from the expected 50 percent is significant. During the wet season, 39 percent of observed daylight resting (20 percent of all resting) occurred between 1300 and 1600 hours, that is, the 3-hour period with the highest observed frequency of resting behavior started and finished 1 hour later, relative to the dry season equivalent. Between 0600 and 0900 hours, 13 percent of observed daylight (7 percent of all) resting took place, which is over $2\frac{1}{2}$ times more than in the same hours during the dry season, and gives an average (4.3 percent) that approaches the statistically expected value (4.2 percent).

During the night and transition half of the 24-hour cycle, resting oc-

a

b

Figure 6.14. Herd resting in the open, (a) at the start of the rainy season, accompanied by numerous cattle egrets (*Bubulcus ibis*), and (b) in the dry season. In (b) note that the six standing individuals are facing in different directions.

curred relatively infrequently from 0500 to 0600 and from 1800 to 1900 hours, accounting for about 3 and 5 percent, respectively (together about 4 percent of all resting). In the remaining 10 hours of that half of the diurnal cycle (1900 to 0500), however, observed resting periods were fairly evenly distributed. These data are shown in Table 6.7.

Choice of resting location

Except in the extreme heat, buffaloes prefer to rest in the open and not in woodland or dense bush. The main reason for moving into wood-

The African buffalo

Table 6.7. *Distribution of resting in S. c.* caffer *throughout the 24-hour cycle (dry and wet seasons) based on data from Zambia and Zimbabwe*

Season	Time in hours (24-hour clock)											
	Day resting (0600–1800 hr)											
Hourly interval:	06–07	07–08	08–09	09–10	10–11	11–12	12–13	13–14	14–15	15–16	16–17	17–18
Dry	53% of all resting											
		5% of DR = 3% of AR					42% of DR = 22% of AR					
Wet	51% of all resting											
		13% of DR = 7% of AR					39% of DR = 20% of AR					

	Night and transition resting (1800–0600 hr)											
Hourly interval:	18–19	19–20	20–21	21–22	22–23	23–24	24–01	01–02	02–03	03–04	04–05	05–06
Dry	47% of all resting											
	4% of NR & TR = 2% of AR			13% NR & TR = 6% of AR			46% of NR & TR = 22% of AR					4% of NR & TR = 2% AR
Wet	49% of all resting											
	5% of NR & TR = 2% of AR											3% of NR & TR = 2% of AR

	Duration of resting periods		
	Observed minimum	Observed maximum	Average
Dry	0 hr 12 min	5 hr 55 min	1 hr 50 min
Wet	0 hr 45 min	3 hr 32 min	2 hr 6 min

Key: DR = day resting; AR = all resting; NR = night resting; TR = transition resting.

land or dense bush for resting is concealment against human and, to a lesser degree, lion predation. My observations lead me to believe that the preference to rest in the open is general throughout the range of the *caffer* subspecies, although in many regions the intensive human

Figure 6.15. A tired herd resting in concealment after many hours of evading lions. Note the open grassland, providing good visibility, on both sides of the small thicket.

threat caused a superficial modification of this pattern. There are several reasons why buffaloes prefer to rest in the open. They are primarily grazers and normally prefer open grassland while feeding. A move to another place for resting obviously entails an expenditure of energy and time that cannot be justified by grazing requirements, especially because immediately after a rest buffaloes normally resume grazing. Although concealment in woodland or dense bush may be a response to predation, a clear field of view is sought even more diligently by buffaloes going to rest in the proximity of predators (Figure 6.15). Furthermore, bothersome flies, for example, tsetse (genus *Glossina*) and others, are much more numerous in woodland and dense bush than in open grassland, at least during the dry season, partly because of an ecological preference for the denser vegetation and partly to avoid winds that interfere with their flight. Heat must be unusually intense before buffaloes will seek shelter from it in deep shade. They do make use, however, of situations that tend to minimize the effect of heat in the open. Buffaloes frequently capitalize on evaporation to stay relatively cool: Although they almost never rest on soggy ground, unless they are wallowing, they carefully select ground with a slightly higher than normal moisture content during the dry season, if this is available. Some individuals are almost pedantic in this and they may spend considerable time searching for such spots, usually peripheral to waterholes or in hollows. When I have tried such places in the heat of the day I noted the cooling effect, as compared to completely dry places. Resting in the open in areas with easily accessible woodland was most frequently observed and photographed in western Zambia, but the most telling example occurred in Kenya, at the southwest extremity of the Loita hills, near Entasekera, in 1966–67. A herd of more than 400 buffaloes rested in the open when hunters, of the sportsman variety, were not detected, but moved into woodland to rest at the signs of a hunting party in the area or of Maasai herdsmen.

Woodland and dense bush are more often used for resting during the wet season. In that season, much open grassland is soggy while woodlands, and even patches of ground surrounding isolated trees in open country, are slightly higher and drier. Woodland and dense bush offer more food during the wet than during the dry season. In addition, when the winds drop during the passing of the intertropical front, coincident with luxuriant growth of the grasses, tsetse (*Glossina* spp.) and other flies are nearly as numerous in the open as in woodlands and cannot be avoided by shifting from one type of cover to the other. These behaviors are not affected by group size.

Posture during resting

Individuals may lie down or remain standing during a resting period. They may be inactive, ruminate, or graze at a token rate. During major rest in relatively safe localities, all individuals composing a herd may lie down one by one and remain lying down for as long as an hour. More commonly, a small proportion of individuals remains standing, most of them facing outward from the middle and periodically adopting the alert stance. Normally, these individuals later lie down, while others stand up. Many of the animals that are lying down, especially those near the edges of the resting herd, also face outward, occasionally become alert, and scan the area in front of them. Both sexes participate in this behavior, which is occasionally displayed by juveniles and fairly often by subadults.

The typical body attitude of buffaloes lying down is identical to that adopted by domestic cattle, forelimbs tucked under the body and hind limbs both to one side. Individual buffaloes show a preference for placing their hind legs on either one side or the other, as do domestic cattle (Fraser, 1974). An individual that is ruminating or at the alert holds its head well up, the nose pointing anywhere between 5° and 40° below the horizontal. The head of an individual at full rest tends to be held low or reposes on some part of a neighbor's body, since resting buffaloes which are linked by close social ties usually form very tightly packed clusters (see Figure 6.16). Body contact may

Figure 6.16. A bachelor club resting on the Lufupa floodplain. (Zambia.)

be important as a means of imparting a feeling of security. In April 1973 in Zambia I rescued a young male calf that had been bogged down in a waterhole and abandoned by the herd. At night I was forced to keep him inside the building I occupied, for fear of predators. The nights were not cold at that time. The calf refused to settle down, vocalizing frequently and slipping about on the floor, unless I placed myself next to him in direct bodily contact, when he would immediately calm down and fall silent. Resting in groups with bodies touching is prominent in other species, notably hippopotamus and domestic swine. Buffaloes of all ages may lie stretched on one side, with the underside of the muzzle resting on the ground or with the head sideways, as horn shape permits.

A state of torpidity that resembles, or actually is, sleep occurs in buffaloes of all ages, but was most often observed in nonadults and old individuals. The animals were standing or lying down, more often the latter. Infants and juveniles may resemble a sleeping dog (Figure 6.17a,b). The eyes are closed, the rate of ear twitching reduced, and other movement absent. The state terminates abruptly and is often followed by vigorous cud chewing. The question as to whether this is true sleep obviously cannot be answered on the basis of visual observations alone, although in the case of calves I do not believe there can be much doubt. Buffalo resting postures are rather highly standardized, with but few variations.

a

b

Figure 6.17. (a) A calf asleep in a characteristic recumbent posture. (b) right bottom corner: a calf asleep. Note the tongue in the suckling (sealing and pressing) position.

Ruminating

Ruminating and resting within a herd occur largely at the same time. Some ruminating can be observed practically at any time in a resting herd, and commonly the majority of its members ruminates simultaneously. The African buffalo most often ruminates lying down, slightly less often standing, and occasionally walking. It resembles in these respects domestic cattle (Fraser, 1974; Hafez, 1975). Buffaloes apparently try to avoid ruminating when they walk but are sometimes constrained to do so by their wish to remain with their group instead of lagging behind. Rumination while walking was most often observed in large groups, perhaps because the larger the group the higher is the probability of some individuals still ruminating when the majority has completed this function and begins to move. Bachelor males in small groups, who function as independent individuals to a greater extent than large-group members, are seldom seen ruminating on the move. Rumination while walking can also be seen when individuals change their location within a resting herd. African buffaloes dislike lying down on soggy ground and therefore tend to ruminate while standing more often in the wet season.

Ruminating was observed directly as the main object of study. However, on many occasions ruminating was also recorded during observations of other behaviors. Numerous attempts were made to obtain continuous records of ruminating in selected individuals over long periods, preferably 24 hours or longer. In most cases the individual under observation was lost from view after only a few hours, which created a gap in the record or terminated it altogether. Moreover, ruminating is affected by the buffalo's state of stress, and therefore observations lost their value if an individual became aware of the observer's interest and became tense in the observer's presence. One may speculate that the cessation of ruminating in buffaloes on becoming tense is an analogous behavior to the cessation of stomach rumbling in resting elephants in similar circumstances. The amount and diurnal distribution of ruminating were successfully recorded in three adults, one at a time, for a total of 69 hours. The animals were: (1) a young adult female belonging to a large herd, which was kept in sight for an uninterrupted period of 30 hours (Zambia, August 1977); (2) another adult female, also belonging to a large herd, observed uninterruptedly for 15 daylight and transition hours (Zambia, April 1973); and (3) a mature bachelor male, one of a group of three, observed continuously during 24 hours (Zambia, June 1974). The length of individual ruminating bouts varied from 2 to 80 minutes. Sometimes a single bout continued uninterruptedly in more than one posture, although very often it stopped when the posture was changed, for a longer time than the normal 4 to 5 seconds required

Table 6.8. *Percentage of time spent ruminating based on observations of three adult (two female and one male) S. c. caffer, Zambia*

Animal	Total observation (min)	Total ruminating (min)	Time ruminating (%)	Ruminating postures (% of total)		
				Recumbent	Standing	Walking
1	1,800	666	37.0	62.0	26.0	12.0
2	900	258	28.7	60.5	34.1	5.4
3	1,440	564	39.2	62.2	35.8	2.0
Combined	4,140	1,488	35.9	61.8	31.1	7.1

to change boluses. Observed recumbent ruminating periods ranged from 8 to 80 minutes, averaging 43.6 minutes. Periods of standing rumination ranged from 2 to 51 minutes, with an average of 14.4 minutes. Rumination while walking occurred in periods of 1 to 10 minutes, averaging 4.3 minutes, and was usually continuous with standing rumination. On the average, there occurred 21.6 minutes of ruminating per hour. Ruminating can occur at any time and has been correlated by Sinclair (1977) in the Serengeti with grazing periods that had occurred 12 hours earlier in the wet season and 2 to 9 hours earlier in the dry season. The timing of rumination in the three individuals observed by me is shown in Fig. 6.4. Other ruminating data appear in Table 6.8.

Chewing, connected with ruminating, was repeatedly observed in five bachelor males for periods lasting between 10 to 50 minutes (Zambia, 1972). Chewing consisted of several series of 33 to 42 chews, at a rate of slightly over 1 second per chew, the last three more rapid, followed by a break of 4 to 5 seconds. Chewing was also observed in captive buffaloes of the *caffer* subspecies at the Antwerpen Zoological Gardens, Belgium (1976). One of these animals, a mature male, displayed a similar chewing pattern to that of the free Zambian buffaloes. It chewed (recumbent) in bouts lasting 6 to 7 minutes each, and 33 to 44 chews per bolus, at rates of 1.1 to 1.5 seconds per chew. However, another adult male and an adult female of the *caffer* subspecies displayed a different chewing pattern: The duration of bouts varied between 4 and 6½ minutes, while chews per bolus ranged from 5 to 31, at the rate of about 1.3 seconds per chew, standing or recumbent. All three captive buffaloes were eating exactly the same food on the same days.

My Zambian data on bolus chewing compare fairly well with Sinclair's (1977) from the Serengeti. Sinclair's counts of chews per bolus range from 41 to 47 approximately, as compared to my range of 33 to 42 for Zambia. A comparison of all these data indicates that the number of chews per bolus can range in *S. c. caffer* adults at least from 5 to 47. While the chewing rate observed in wild Zambian buffaloes was also observed in a zoo animal, I have no confirmation that the much fewer chews per bolus observed in some other zoo buffaloes also occur in the wild.

My observations agree with Sinclair's (1977, p. 92) that nonadults chew faster than mature adults, but unlike Sinclair it appears to me that such a trend is to be expected. This trend may be favored by a probable change in bolus size during the early years and probable substantial changes in the ratio of bolus size to chewing capacity in growing buffaloes. With maturation, the jaws increase in length, placing the cheek teeth further from the condyle and thus increasing the moment arm. The lateral excursion during the grinding stroke of the lower jaw during mastication increases and the musculature operating the mandible becomes more powerful. Thus in the adult buffalo the food being chewed is located farther from the hinge of the jaws than in the juvenile, while the sideways grinding stroke of the lower jaw is longer. Consequently, a slower, more powerful, but longer grinding stroke would provide the most economical use of energy for chewing in adults. The juvenile must chew faster than the adult to achieve similar trituration because increased frequency of chewing must compensate for less powerful musculature and shorter stroke.

Tooth wear and the consequent decrease in chewing effectiveness may be one reason why old males lag behind their herds and eventually adopt a separate existence, for this permits them to graze more selectively and longer than would be possible while keeping up with the herd. Old females do not follow the same pattern probably because their herding instinct is stronger than it is in old males and keeps them more closely bound to their herd. If serious tooth wear, inevitable in old bovids, could not be compensated for in some way, such as the one suggested above, all old individuals could be expected to show signs of malnutrition. However, in my experience, a large proportion of old isolated bachelor males maintain very good condition, as compared with that of local herds at the same time. Old individuals within herds often show poorer condition than the general level within the herd. Old females may have a higher mortality than old males because the former tend to remain within herds, whereas the latter are often solitary or in bachelor groups for their last years.

Summary of resting and ruminating behavior

Resting and ruminating often occur together, although ruminating may take place at other times. In the absence of intensive human threat and extreme heat, *S. c. caffer* prefers to rest in the open. Resting can occur around the clock, is usual in the afternoon, but is rare in the transition from night and day. The main difference in the dry- and wet-season resting patterns appears to be a more even distribution of resting during the latter. A recumbent posture is preferred for ruminating. However, contact with soggy ground is avoided and ruminating occurs while standing if the ground is wet. Ruminating while walking is rare. *S. c. caffer* appears to have a preferred rate of bolus chewing, ranging from 33 to 47 chews per bolus, at slightly more than 1 second per chew, but mastication of less than 31 chews per bolus occurs in zoo animals. Nonadults tend to chew more rapidly than adults. The separate existence of some old animals (mainly males) may be due in part to a need for a more selective, and perhaps longer, grazing regimen to compensate for tooth wear than is possible while keeping pace in a herd.

Eliminating

Marked similarities exist between the eliminative behaviors of *S. c. caffer* and domestic cattle of the genus *Bos*, described by Hafez and Bouissou (Hafez, 1975). Like domestic cattle, *S. c. caffer* defecates anywhere and does not markedly avoid walking or lying down in its own or a conspecific's excreta. As in domestic cattle, the greatest amounts of excreta occur in places where buffaloes remain longest, and therefore coincide with resting sites. Frequently observed hoof imprints in fresh dung as well as dung pats rolled out by a recument buffalo's body are direct evidence of this behavior even after a herd's departure. This is in contrast with the eliminative behavior of feral water buffaloes (*Bubalus bubalis*) in Australia, which make use of a "dung heap," near but separate from their camping (i.e., overnight resting) spot (Tulloch, 1978).

   Defecation may occur either while standing or walking. The stationary defecating posture is like that described by Hafez and Bouissou for domestic cattle (Hafez, 1975, p. 235): "Typically, a special stance is assumed: the base of the tail is raised and arched away from the body, the hind legs are placed slightly forward and apart, and the back is arched." The walking posture while defecating is similar, as illustrated in Figure 6.18a. Females have been observed urinating only in the standing posture, which is spread-legged and somewhat similar to those adopted when defecating or expecting to be mounted by a male (Figure 6.18b). Males urinate standing or walking, which again resembles domestic cattle, but unlike males of domestic cattle, which urinate slowly and under little pressure (Hafez, 1975), *S. c. caffer* may expel urine under considerable pressure, which may cause the penis to vibrate rigidly.

   The consistency of buffalo feces varies with water intake (through

a

b

Figure 6.18. (a) A buffalo eliminating while walking. (b) A young female in the urinating/defecating posture, which resembles the stance assumed by estrous females expecting to be mounted by tending males.

drinking and ingestion of moist food) and also with the texture of the graze or browse. Feces range from splashy, after abundant and frequent water intake, to a mass of compressed subovoid lumps when little water has been ingested.

The number of defecations during a 24-hour period is variable, both in number and distribution. Herd members defecated mainly either when stationary, usually at a resting place, or when walking, and no correlation was discovered between the preference to do the one or the other, with seasons, type of pasture, or water distribution. Spooring of four solitary adult males over ground covered by them in 24 hours gave the following feces counts:

dry season, Mabalauta, Zimbabwe   4
dry season, Chunga, Zambia   8
wet season, Chunga, Zambia   7
end of wet season, Busanga, Zambia   11

Urinating often occurs, in different individuals, at the same time as defecating. As a final parallel with domestic cattle, *S. c. caffer* does not appear to attach any social significance to excreta (for cattle see Schloeth, 1958).

7

Hierarchy, status, and individual specialization

Most of the factors that are generally considered to enter into the determination of an ungulate's status within its social order appear to play a part in the *S. c. caffer* hierarchy. Thus individual rank of adults is determined by sex, age, and physical type, which commonly means superior size and strength.

Hierarchies in small buffalo groups in which all the individuals were identified always were linear, either in the exact sense or with minor complexities, such as occasional triads in female sequences. Consequently, attempts to determine hierarchies in larger sets, in which not all the individuals could always be identified, were based on the assumption that here also linear or linear-tending orders were present. The proportion of fully recognizable animals in such sets usually exceeded 50 percent. When a number of interactions had been recorded between pairs of recognizable individuals and between pairs consisting of one recognizable and one unrecognizable individual, the recognizable individuals were arranged according to their apparent relative rank. The unrecognizable individuals, which nonetheless were often divisible into two or more distinct classes, were then fitted among the fully identifiable individuals in the best possible way to accommodate a linear order. There were no cases where the available data made a linear-tending order impossible. The few contradictions to a simple linear order could usually be explained by introducing a triad sequence. Repeated open aggression is rare in *S. c. caffer*; hence moderate- and low-intensity agonistic interactions were recorded. Preference was given to pairs of interactions between the same two individuals, actively displaying reinforcing dominant and submissive roles to each other, especially if the interactions were of low intensity.

There is little doubt that all adult males of *S. c. caffer* are dominant to all nonadults and probably to all breeding females. The only females that were observed to be dominant to some adult males had strong secondary male traits. Among 37 such females observed in Kenya (13), Tanzania (5), Zambia (14), and Zimbabwe (5), only 1 was with a calf that appeared to be her own, and I suspect that such females are usually sterile. In the total buffalo population examined, I would estimate the proportion of such females at about 0.25 percent (minimum). Eighteen of the 37 were observed in 108 nonviolent agonistic encounters with adults and subadults. In 75 percent of the cases the females with secondary male characteristics were the aggressors. Of the 108 encounters, 55 were with other females, including 5 subadults and 6 pregnant individuals. Of the female–female encounters, 89 percent were won by those females with strong secondary male traits. This included 3 of the encounters with pregnant and all the encounters with subadult females. Of the 53 observed encounters with adult or subadult males, 47 percent were won by the females. Of these encounters 38 were with adult males, and 9 of these (24 percent) were won by the females. Thus females with strong secondary male characteristics may occupy a special position in the hierarchy, almost independent of sex.

Aggressiveness in these atypical females appears to be higher on the average than aggressiveness in bachelor males, when expressed as the number of agonistic interactions per individual per hour. Agonistic encounters between these females and adults or subadults of either sex ranged from 0.78 to 2.94, average 1.76 per individual per hour, over a total of 66 hours. If only the encounters between adults are considered, the range is 0.78 to 2.67, average 1.53 per individual per hour. The equivalent for 51 bachelor males in eight groups in which a total of 151 interactions was observed over 77.5 hours ranges from 0.17 to 0.79, average 0.39 per individual per hour.

Male hierarchy

The dominance order was linear in eight sets (or segments of groups) of bachelor males, numbering between 4 and 11 individuals, all of which were recognizable. My own observations from Kenya, Zambia, and Zimbabwe, Grimsdell's (1969) from Uganda, and Sinclair's (1977) from northern Tanzania suggest that such situations generally prevail. However, partnerships between pairs of males may introduce complexities. I have observed two instances of a subordinate male "borrowing" status from his partner. One involved a group of 4 males (A, B, C, and D), related by a simple linear hierarchy on the first day of their separate association after breaking away from a larger group. Two days later, males B and D were observed keeping close company. From that day, while male A remained dominant to all and B was dominant to both C and D, C started to behave subordinately to D. D had thus apparently "borrowed" status from B, to rise above C. Another case involved 7 males that were with a herd, forming part of a bachelor club of 11 individuals, which I was able to rank linearly from A to K. The club was broken up when males C and E to J separated from the herd. These 7 individuals retained their relative ranking. Within a day, males C and H began to keep close company, often slightly separate from the other five individuals. Subsequently, while C displayed the expected dominant behavior toward H, E and F displayed subordinate behavior toward H. H also won a passive encounter with E. Both examples were observed in Zambia, in 1974 and 1977, in the dry season.

The dominance order may be tested by sparring bouts between pairs of usually closely matched individuals, especially in the subadult and young adult segments where the order of dominance is still fluid. While sparring is a prominent behavior of immature individuals, I have also observed it frequently between fully mature males.

Herd males (as opposed to bachelor males) are not normally displaced by direct confrontations with bachelor challengers but appear to leave mixed groups on their own initiative, to join a bachelor club or live by themselves. When a bachelor male moves in among the females, he is either confronted by a herd male, in which case he is easily outfaced or defeated and withdraws, or he is not confronted and remains in the mixed group with the females. Apparently this kind of nonconfrontation is a form of agonistic interaction where the bachelor is the winner. The freshly accepted male is then treated by the females as a preferred individual but not a leader (Chapter 4).

In most herds there is at least one male with very high status, although there may be several. When there are several such males, mild agonistic encounters occur among them, defining a linear hierarchy. I have seldom observed these high-status males engaged in agonistic interactions with males of lower status. Such few interactions as I did observe were initiated, and immediately won, by the high-status males. The high-status males can be violent on such occasions and may use their horns to gore as well as buffet.

High-status males are often found in small groups or individually at a few tens of meters from the herd. When the herd is on the move, it is common for them to trail behind, but occasionally one or all of them take the lead. When some females are in estrus, these males are usually located within the herd, each near an estrous female. When a herd approaches a waterhole, the top male and other high-status males may delay their arrival until all the others have drunk and moved away. They then drink, frequently wallow, and may horn and toss

clumps of grass and mud at the edge of the waterhole before following
the herd (see Chapters 8 and 12). If the watering place consists of
several waterholes or is a large body of water such as a river or a lake,
the high-status males do not lag behind the rest of the herd but find a
spot at a little distance from the others. They do not normally wallow
in open water, however, possibly as a precaution against crocodile
predation. Although the top males do not usually guide a buffalo herd
on the move, as a rule it is they that initiate intensive grazing or
movement at the end of a resting period, by their own activity and
vocalization.

These high-status males are not the only ones that serve females in
estrus. Herd males of lower status may copulate with females but are
often displaced by one of the high-status males toward the peak of
estrus, at ovulation. This is described by Sinclair (1977) for the Serengeti buffaloes, and is a widespread behavior for all buffaloes, even
when in small groups.

The main function of the top males appears to be the insemination of
a high percentage of the females, a situation encountered in various
other gregarious species (Lorenz, 1966). They are the only males that
commonly transgress basic-herd boundaries although, when herds
split up, each male tends to adhere to one particular basic herd, if not a
bachelor club.

Old bulls are gradually phased out of the hierarchy. Sometime after
his tenth year the S. c. caffer male undergoes a decline in reproductive
ability (Grimsdell, 1969) and reaches a critical degree of tooth wear
which may markedly impede his grazing efficiency, as compared to the
younger animals making up the bulk of the herd. These changes are
often accompanied by the development of various chronic ailments and
a tendency to lose weight and condition, all of which are corollaries of
status. The rates at which individuals are affected thus vary widely and
consequently males respond variously to the onset of old age. Those
old males that are able to feed at the same rate as herds and therefore
need not separate in order to survive may stay with them or attach
themselves to bachelor clubs. Their status becomes ambiguous: They
are frequently given the right of way and may keep company with
high-status males but they are likely to be horned if they move in on
females, as observed on three occasions. However, they do not normally attempt this. In time, as their strength wanes, they lag behind
and eventually lose contact with the herd.

Other old males leave the herds earlier. They do so singly or in
partnerships of two or three, occasionally four or five, and do not
rejoin the herds again except as the result of a chance meeting, when
they tend to attach themselves to invalid groups. In most partnerships
of old bulls that I have observed a simple dominance order existed. In

pairs, the dominant-to-subordinate relationship was usually very apparent, manifested by numerous low-intensity interactions.

Female hierarchy

Female hierarchies, distinct from the male, are somewhat more complex. At the basic-herd level I concluded that they are linear-tending, with more frequent complexities of the triad type than in the male. During 1965 I observed and plotted the social order in as much of the Lolgorien herd as I could. Three separate female dominance chains appeared, corresponding to groups A, B, and C (see Chapter 4). Both seniority and physical type seemed to play a part in ranking within groups. In encounters between individuals from different groups, age did not seem to give an advantage. Figure 7.1 depicts schematically the relationships within the Lolgorien herd at a time of my contacts with it.

In group A there was a high-status elderly female (PE) without a calf, normally well forward in the group, with whom a similar-looking younger female (MD) kept company. The second female was suckling an infant and was accompanied by a female juvenile, apparently a daughter. Two other juveniles, 1 male and 1 female, usually stayed with them. Closely associated with the 2 females and their juveniles was a set of 10 females with 4 infants (1 male and 3 female), 3 juveniles (1 male and 2 female), and 4 subadults (1 male and 3 female). The subadults tended to cluster together. The most dominant of these 10 females (PV) was a robust cow of advanced maturity, with a slightly different appearance than the first 2 cows, but 1 of 4 animals that strongly resembled one another. Usually to the rear of this set were 5 females, of which 3 had infants. A male juvenile was associated with one of these females. Two female subadults, usually together, completed the set. A single female tended to remain at the end of the group. Five males, 4 adults, and 1 subadult were usually mingled with group A. The subadult tended to trail one of the adult males.

Fifty-six interactions between adult females suggested a linear-tending order, with triads linking females assigned numbers 1, 2, 3 and 15, 16, 17. As noted above, the first 3 females at the top of the order were also coded PE, MD, and PV, respectively. The average number of encounters per individual per hour was 0.41.

The first-described high-status elderly female, PE, was very energetic; her somewhat younger companion, MD, was very much of the same physical type and could be her daughter. Number 3 female, PV, was a powerful animal of a slightly different physical type from PE and MD, a type shared with three other females from which PV differed only in the degree of robustness and apparent greater seniority. This female was observed winning some encounters with PE, while losing

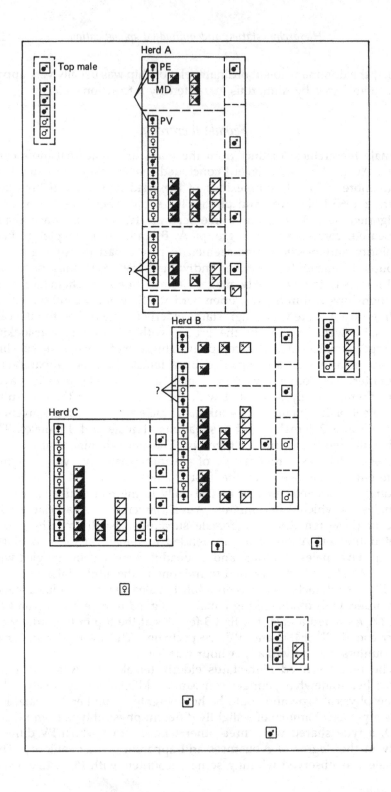

them to MD. Female number 1, PE, was shot. Seven days later, MD with her infant and still accompanied by the female juvenile was observed keeping a position slightly behind and well to the side of PV, who was definitely in front, closely followed by the 3 cows that resembled her and the others that were included in the previously mentioned set of 10 adult females. Although MD kept drifting toward PV, she did not make her way forcefully to PV's side, as was her way with PE, but remained well to one side. I did not observe MD in an obvious encounter with PV at that stage, but MD's general behavior appeared to be subordinate. It was impossible to determine exactly the subsequent position of MD with group A, as I lacked the time for the necessary observations, but she appeared to have fitted into one of the first 11 positions, possibly well forward. Unfortunately, there is no proof that MD was PE's daughter, despite physical resemblence and the close association between those two females. However, regardless of the genetic relationship, it appears that MD acquired status that was

Figure 7.1. Schematic of the Lolgorien buffalo herd, when it numbered 139, in 1965. The configuration of the component groupings is a probable one, based on field observation, when going to pasture (movement toward the top of the page). The adult females in each of the three basic herds, initially called groups A, B, and C (Chapter 4), are schematically aligned, from top to bottom, according to the most probable order of rank as deduced from agonistic interactions between individuals which were fully identifiable (solid circle in sex symbol). The adults marked not as fully identifiable (open circle in sex symbol) were confused in some cases with only one other individual, and therefore educated guesses as to which was which were often possible. There is no attempt in the diagram to connect individual nonadults to their mothers. However, the basic herds are subdivided into identifiable "sets" of adults, and nonadults staying with each set are shown separately. Remarks regarding some mother–child and sibling links occur in the text (Chapter 7). The three females referred to in the text as PE, MD, and PV are identified here. The female with strong secondary male traits in basic herd (group) A, herself childless, herded at one time several nonadults from her own and other sets (an entity referred to in the text as a juvenile club). Several adult males remained most of the time attached to basic herds. They are depicted as members of larger sets or as forming sets of their own. The number and individual membership of bachelor clubs was rather fluid – hence the broken line enclosing them. Legend:

Solid lines:
  Enclose basic herds A, B, and C
  Enclose entire herd

Broken lines:
  Enclose bachelor clubs
  Separate units within basic herds

 Triad   ?  Probable triad

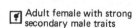 Adult female
Adult female fully identified

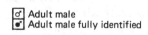 Adult female with strong
secondary male traits

Adult male
Adult male fully identified

 Subadult male
Subadult female

 Juvenile male
Juvenile female

 Infant male
Infant female

"borrowed" from PE and had to revise her rank downward on PE's death, allowing PV to slip into the number-1 position. The "borrowed" status of MD could therefore have been the reason for the triad.

In group B, two females of advanced maturity occupied the top positions. Number 1 had no young offspring but a female subadult stayed near her. Number 2 was suckling a male infant and was followed around by a juvenile female. Forming a set with them were 2 more cows, one with a female infant. The childless one, number 4, was part of a possible triad, completed by the female assigned the number 6 and one of 3 similar females which were placed in positions 5, 9, and 14 that I tended to confuse. If the questionable individual was 5, then there was a classical triad. Number 4 adhered to the top 3 females. Numbers 5 to 10 formed a somewhat separate set, although the top 2 may have been forming the supposed triad with number 4. There were with these females 4 infants (2 female and 2 male), 2 juveniles, 1 of each sex, and 4 subadults (3 female and 1 male). A mature male integrated into this set. The 5 remaining adult females kept more or less together, with their 3 infants (2 female and 1 male), 1 male juvenile, and 1 subadult female. Besides the 1 male already mentioned, 4 others mingled with group B. A linear-tending order was assumed on the basis of 46 interactions between adult cows. The average number of observed encounters was 0.38 per individual per hour.

In group C there was no hint of any subdivision: It was a close-knit cluster of 11 females with 7 infants (4 males and 3 females), 2 juveniles (1 male and 1 female), and 4 subadults (3 females and 1 male). Two subadult females kept close company. Two mature males were commonly to be found among these females and 4 other males tended to accompany group C, weaving in and out of it. There were two interactions incompatible with a simple linear order, and these could be explained as minor complexities of the triad type. Thirty-two interactions between female adults were recorded, and the average number of interactions per individual per hour was 0.48.

Besides the three groups that have just been described, there were 17 adult males, 6 subadult males, and 3 adult females that did not have any group affiliations. They tended to divide into 3 clusters and a few unattached individuals. The top male and 5 high-status males often formed a cluster of their own.

Interactions between adult females within each of the three basic herds were observed on separate days. Concurrently, interactions between members of the observed basic herd with those of the other basic herds were recorded. All such interactions numbered 78, among them 53 of females with adults of either sex, including 30 with other adult females. Estimates of average numbers of interactions per individual per hour give a figure of 0.03. Even if it is assumed that I missed

75 percent of this type of encounter, the figure would amount only to 0.13, prorated on what I observed within the basic herds. Thus it may be concluded that interactions between basic herds are considerably fewer than those within basic herds.

Similar dominance chains were determined for the 7 adult females of a small herd of 19 individuals observed east of the Ntemwa river in the Kafue National Park in October 1977 and for a part of a herd in the Busanga area comprising 9 adult females in July 1974, both in Zambia. The frequencies of female–female interactions average 0.37 and 0.52 per individual per hour for these two groups, respectively, which is comparable with the 0.41, 0.38, and 0.48 observed in the Lolgorien basic herds.

Individual specialization

Herds are usually guided along their routes of march by one or more individuals that are not necessarily at the top of their hierarchy. They can be of either sex and of any age between subadulthood and advanced maturity. I call such animals "pathfinders." All adult pathfinders whose ranking was determined belonged in the upper half of their hierarchies, and a few of them were identified as high-status individuals. There were individuals that acted as pathfinders in all the buffalo groups that I observed. A buffalo herd may adhere for a long time to an established route whose length may be several tens of kilometers. Individuals tend to specialize as pathfinders for particular portions of a route, and most times a herd reaches a certain locality a particular individual crowds forward and takes over the leadership, without opposition. The pathfinders' function is unevenly shared among them and only one or two may do most of the leading, but there is no rule in this respect from one herd to another. If the top bull functions as a pathfinder, his overall status is very high; he is never impeded in any way and becomes in effect "the lead bull." Such a top bull automatically walks first and it is difficult if not impossible for him to transfer leadership to another individual while the herd is advancing, as is commonly done between pathfinders of lower status: When the leading top bull stops or hesitates, all the rest stop or markedly decelerate, whereas if the herd is headed by a lower-status buffalo this animal is often relieved by one from behind if it hesitates, and the herd keeps advancing.

The proportion of females with strong secondary male traits is high among pathfinders. Of 57 individuals from various herds which were identified as pathfinders, 6 were such masculine cows. This constitutes 10.5 percent, whereas my estimate of the proportion of these females in the total population is 0.25 percent, a difference factor of 42. It was noted (Chapter 4) that 2 out of 5 females observed leading groups of nonadults also belonged to this type. There is here a suggestion that such females are driven to compensate for their reproductive malfunction or sterility by adopting other roles.

One of the more interesting aspects of pathfinder behavior is the way in which the function devolves on the same individuals, on different occasions that are widely spaced in time. I will describe one particular case of pathfinder behavior in which this element was prominent. In the Busanga area (western Zambia) in July 1974 I had observed a basic herd, normally part of the Lushimba herd, when it was separated as a result of lion attack and functioned on its own for some days afterward. It was then largely guided by a young adult female that I had no difficulty in identifying. When this basic herd was included in the Lushimba herd, this female was not among the herd's usual pathfinders but moved within her own group, without any apparent special function. I had identified this female several times afterward, between then and 1977. On July 2, 1977 at about 0830 hours, I chanced to drive my Land-Rover near the Lushimba herd as it was watering at some pools by the motor track. The herd, which then numbered over 400, took fright at the sight of the Land-Rover and started to flee westward across the track, crossing ahead of my vehicle. Most of the herd crossed, but two basic herds (about 100 individuals) remained east of the track, "cut off" by the slowly advancing Land-Rover. I drove past the crossing place and parked, while the two basic herds stood behind a large termite mound, the main part of the herd on the opposite side of the track retreating in the meantime westward into woodland. The two basic herds were in a congested mass, impossible to separate by eye without some individual recognition. After 3 minutes of waiting, the above-mentioned female (now about 9 years old) broke out of the middle of her group and moved purposefully to the front, about 15 meters in advance, westward, of all the others, halted in an alert posture and carefully probed the air, lowered her head, and, sniffing the grass, advanced at a mincing run some five steps, then probed the air again. At this time (i.e., after some 2 minutes) a second female broke out of the other basic herd and positioned herself 3 meters behind the first, well in advance of the remaining buffaloes. She moved, not in time with the first female but maintaining approximately the same distance behind her and behaving similarly, that is, repeatedly probing the air, advancing a few paces, and sniffing near ground level. The two basic herds (which contained no fewer than 20 males) followed the 2 females in a body, without showing signs of alarm or taking any individual precautions that I noticed, thus displaying submission to this leadership. The buffaloes crossed the track and eventually joined the bulk of the herd. Whenever I subsequently saw the basic herds they were always within the greater herd and the females in question were not functioning as pathfinders.

Alertness against possible danger is more pronounced in some adult females than in others, as indicated by four series of observations of identifiable females in Kenya, Zambia, and Zimbabwe. When a female that could be identified was seen to assume the alert posture several times at relatively short intervals, two or three other nearby females were noted as "controls" and all of them were observed for 2 hours. The length of time and the number of occasions each female assumed the alert (scrutinizing) stance was recorded. The results for sets of females for which 2-hour observation periods could be repeated three times were retained. These observations indicate that some females spend consistently more time scrutinizing the herd's surroundings, and adopt the alert stance more often, than do other females (Figure 7.2). A female's level of alertness apparently does not correspond to her position within the hierarchy. Similar observations on males failed to produce consistent results.

An isolated bachelor club

All of the features that bachelor clubs have in common were illustrated by the activities of the largest one that I observed. It was encountered in Zambia during October 1971 on the Busanga flats near the Lufupa drainage, at the end of the dry season. It numbered 38 individuals: 33 vigorous adult males, 3 males somewhat past their prime, 1 old, and 1 immature male. They were in good condition, and the only defects noted were that one bull lacked the lower portion of the tail and the oldest was unable to hold his head straight or bring it up as high as is normal for a healthy adult.

Twelve individuals were absolutely identifiable, whereas identification of a further 9 could be slightly dubious under poor conditions. The remaining 17 individuals could be identified and followed for a few hours at a time, until lost in the group. Subsequent identification of the same individuals was speculative, although patterns of association with others facilitated identification. The confusion resulted from similarities in conformation, age, and lack of individual characters.

The buffaloes were observed for 168 hours from the first sighting. A total of 127 hours (76 percent of the entire period) was spent on actual observation, which was interrupted only by brief catnaps while sitting up. Throughout the whole 168 hours of observation time no other buffaloes were seen in the same vicinity.

During the 168 hours the males remained together within an area of about 150 square kilometers. An exception occurred when 9 animals wandered apart for several hours. On two occasions the bulls moved

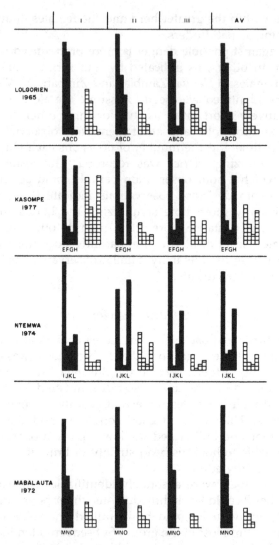

Figure 7.2. Alertness against possible danger appears more pronounced in some adult females than in others. Each observed adult female is denoted by a letter. Each pair of histograms represents a 2-hr observation period (i.e., each individual has been observed three times). Black histograms show the frequency of alert behavior expressed as percentage time spent at the alert during the observation period; white histograms show the frequency of alert behavior expressed as the number of times the alert stance was assumed during the same period. A strong correlation is apparent between particular individuals and degree of alertness, as compared to the other individuals forming a set. All were herd animals.

briefly into woodland, but most of their time was spent in the open grassland of the dry Lufupa floodplain. The bachelor males remained in the same vicinity from their first sighting by me on October 21 until November 7 (S.H. Logsdon, personal communication), that is, for at least 18 days. Browsing was not observed when the group was in woodland, although it could have been missed during the night.

The bulls were not particularly active, and signs of restlessness were uncommon. Most of their moves were strictly local, and neither a marked change in weather (clear and hot to cloudy with wind and drizzle) nor a night encounter with lions caused the bulls to move out of the area in which they were first sighted. Serious fighting did not take place within the bachelor club while it was under observation, but sparring did occur. Seven sparring contests were noted during daylight and two during the night. One of the latter was identified only by sound and was probably the most violent engagement of all. During daylight no other aggressive incidents took place, but it is possible that some relatively silent contests were missed during the night. All the daylight contests occurred either in the morning (two out of seven) or in the late afternoon (five out of seven). In all cases the individuals concerned, as well as the remainder of the club, were relatively inactive previous to the contest.

Although the club was made up of several groups, there were indications of an overall hierarchy within it, topped by one distinctly dominant bull. Four components were distinguished within the bachelor club, comprising groups of 13, 10, and 7, and 6 individuals that usually stayed in threes, pairs, or singles. I designated these groups I, II, and III and "group" IV, respectively. Sometimes one or more "group" IV individuals became attached to one of the true groups. Less frequently members broke away from their groups, remained alone, and then reattached themselves, sometimes to groups other than their own. Such rearrangements, however, did not last. At the end of the next general movement of the club the basic subdivisions were reestablished with few if any irregularities. The groups tended to keep their identity when the club was resting by maintaining some distance from each other, but members of the same group usually touched one another and rested their heads on neighbors' backs.

Group I contained a big middle-aged bull that was dominant to all the remaining 37 animals. Group III contained two high-status individuals, and group II appeared to contain three such individuals, but they were extremely similar and I could not be sure that I did not confuse them occasionally. I was able to determine a simple linear hierarchy in the seven members of group III and in the four top members of group I. The remainder of the bulls appeared to have a linear hierarchy as far as I was able to tell. Interactions between pairs of males from different groups indicated harmony across group boundaries, as well as the probability of an underlying single linear-tending hierarchy for all the animals, with the high-status males of all the groups located near the top. However, irregularities through "borrowing" status from partners, and conceivably of the triad type, were not impossible.

The top status of the dominant bull of group I was apparent throughout the bachelor club. Although group I, including the top bull, rested frequently in the middle of the club, it was always the one to initiate general movement. The typical pattern of transition from total group rest to activity is shown in Figure 6.10. To start general group movement, the top male gradually worked his way through the group, in the direction in which he intended it to move, and eventually ended up about 10 meters or more in the lead. From then on, so long as the top male made an effort to stay ahead, the rest followed. If the top male stopped or slowed down sufficiently to allow any of the others to overtake him, it appeared to serve as a signal that the move was over.

When the club was at rest or grazing, a number of individuals undertook to act as lookouts. Some individuals did this more than others, but there did not appear to be a connection between this behavior and rank. A connection appeared to exist with the individual's location at the time and perhaps a relatively low threshold for the release of alert behavior. However, the top male was almost always one of the lookouts, and there were times when he alone kept watch. On one occasion a Land-Rover, one of three that came near during the survey, frightened the bachelor club. A short time after the car's passage all the buffaloes were again grazing except the top bull, who continued to listen attentively until a moment after the engine sound had become inaudible to me (possibly also the buffalo's limit of hearing?) and only then resumed grazing. On other occasions the top bull was the only one to carefully scrutinize the surroundings at the alarm whistle of a puku antelope (*Kobus vardoni*) while the remainder of the bachelor males continued grazing.

The bachelor club grazed intermittently between 0530 hours and 0830 hours. Between 0830 and 1000 hours the club was normally resting and grazing was at a minimum, with least activity about 0900 hours. At that time none of the males grazed, or only a few did so, sporadically, and it was common for all to lie down. From 1000 hours onward grazing activity increased, to a level that was higher than in the early morning before the resting period but that nonetheless can be described as spasmodic. The most active grazing period during daylight was from 1400 to 1730 hours, when the club grazed almost continuously at moderate to intensive rates. One afternoon was distinctly overcast, with strong wind gusts. On that occasion grazing activity was much lower than normal and sometimes ceased altogether. It was my impression that the wind was the main factor in this change of the grazing pattern (see Chapter 6). After 1730 hours grazing dropped to approximately the same level as in the 1000–1400 hours period, until nightfall at about 1845 hours. During the hours of darkness grazing continued intermittently, with occasional lulls but no observable pattern. Vocalization was very scarce at all times.

During the night of October 22–23, the bachelor club drifted into woodland and soon after dawn it was resting in groups. At that time the dominant male of group III became excited, looking about, smelling the air, and rubbing himself against other individuals in the group. Then he started to walk toward the open grassland, followed by the rest of his group and two bulls from "group" IV that formed a pair nearby at the time. The remainder of the bachelor club did not move. The pace of the nine bulls was a fast walk, the group's dominant bull leading, the others following in a stepped line, *en echelon* , accompanied by six cattle egrets (*Bubulcus ibis*). I went to collect my Land-Rover and rejoined the nine bulls. They walked for over 2 hours, halting only once for a moment in alert stances, and covered about 10 kilometers. They reached a small pond surrounded by green grass near the Lufupa drainage and scattered and grazed for 20 minutes. They resumed their walk until they reached a swampy area in the Lufupa drainage 8 kilometres further on, dispersed, and began to graze. I returned to the main group. In the early evening the seven bulls of group III reappeared. The time was 1749 hours, soon after the intensive grazing period, and all the bulls but one in the main group were lying down. The newcomers stopped to drink some 100 meters from the main group. One chased some egrets out of his way with his head close to the ground. The egrets had flown from the main group to meet the newcomers. Eight minutes later one of the newcomers gave a long bellow, and one of the high-status bulls in the main group, but not the top bull, stood up and looked toward them. When the returning bulls reached the others, members of the main group remained lying down but the newcomers did not lie down. The remaining two bulls, the pair from "group" IV, rejoined the club after dark. A possible explanation for this day-long excursion may be a drive to rejoin the herd. Although the herd was not nearby at the time, the locality to which the nine bulls went was an important grazing and resting area on its circuit. Furthermore, the time of year, that is, the transition from the dry to the rainy season, is one during which isolated males tend to return to herds. The wind was against the bulls and they probably could not know by scent that the herd was absent. When they failed to discover the herd, they gravitated back to the bachelor club. This explanation of the excursion might imply a strong selection factor operative among bachelor males, not only with regard to sexual drive, causing some to go in search of the herd while others do not, but also for the formation of the more sexually vigorous males into separate groups within the bachelor club. Alternatively, only the dominant male of group III may have been driven to search for the herd, and the others may have merely followed his lead.

A noteworthy behavior is the "vetting" of the new arrivals not by the normally vigilant top bull of this bachelor club but by a lesser individual. The top bull's apparent indifference is a typical behavior of individuals that are expert at scrutinizing, and I consider it can be simply explained by the early recognition of an event as comprising no threat.

Cattle egrets (*Bubulcus ibis*) normally joined the bachelors between dawn and 0700 hours, arriving gradually in pairs and small flocks or larger flocks of 20 or 30 birds. They usually left the buffaloes just after sunset, most often in large flocks. The greatest number counted at any one time was 64. From time to time during the day all or some of the egrets left the buffaloes for periods of up to several hours and then returned.

Red-billed oxpeckers (*Buphagus erythorhynchus*) accompanied the bachelors, feeding on parasites that lived on their bodies.

The bachelors came in close contact with blue wildebeest (*Connochaetes taurinus*) and Burchell's zebra (*Equus burchelli*) three times during the survey. On one occasion the buffaloes rested next to the wildebeests; on another a herd of 10 wildebeests came to drink beside the buffaloes; and on a third, the buffaloes rested with some 23 wildebeests and 7 zebras, grazing subsequently next to both species.

During the night of October 23–24, two lions approached the bachelor club on open ground and apparently attempted to make a kill. This unsuccessful attempt was observed at close quarters and although there was no moon, the starlight on the open plain provided reasonable visibility. Details were checked by walking over the area and measuring the distances on the following morning. The description is included here since the behavior may be limited to bachelor clubs.

The bachelor club was grazing in the open by the Lufupa drainage. At about 0300 hours two lions approached but were not sensed by the buffaloes until less than 50 meters from them. On sensing the lions the bulls halted and stopped grazing. Almost immediately afterward they clustered very close together, facing the lions. The lions, one a large male and the other probably an adult female, came to within about 15 meters of the buffaloes, moving slowly around them and watching them. They walked one behind the other, about 5 meters apart. As the lions moved, the buffaloes kept shifting around, so that they were always facing the lions but without markedly changing their ground. The lions walked some 270° clockwise around the group, then halted for a short time and started to retrace their steps counterclockwise. When the lions were part way back around the arc, four bulls advanced in a tight cluster to within some 5 or 7 meters of the lions. The lions halted, faced the buffaloes, then continued to move, and eventually left. When the lions were gone, the bulls resumed grazing, dispersing slowly, but about six remained at the alert for 15 to 20 minutes afterward.

I felt the seriousness of the lions' intent had to be substantiated as far as possible so I followed their probable direction in the morning and discovered a wildebeest kill a few kilometers away. This suggested that the lions were hungry since buffalo (and zebra) are the favorite prey of lions in that area and wildebeests are not favored, as they are in East Africa.

Summary

It appears that linear and linear-tending orders exist within small social units of *S. c. caffer*. The smallest mixed social units observed in the *caffer* subspecies, which I call basic herds, are normally larger than family groups. The female hierarchies are more lasting than male hierarchies but any males that remain together, even only for a few hours, soon establish a usually simple linear dominance chain. Several linear-tending female hierarchies coexist within one herd. Interactions between females of different hierarchies appear to be fewer than those between females belonging to the same hierarchy, and can probably be explained by random encounters of individuals in zones of basic-herd overlap. The herd cannot, however, be simply defined as an aggregate of basic herds, because it possesses a few social bonds, such as the uppermost portion of the male hierarchy that is shared by the entire herd or the occasional acceptance of overall leadership and initiative of individuals. Thus the herd has a loose but real social structure, reinforced by the buffalo's apparent innate tendency to aggregate. A few individuals perform a disproportionately large share of some functions related to herd leadership and protection, and therefore may be considered socially specialized. Such specialization is not directly correlatable with position within dominance chains. Large isolated bachelor clubs may be socially well integrated, with organizations approaching those of herds.

8

Agonistic behaviors

I. Interspecific aggression, threat, and responses

Buffalo–human encounters

Threat displays

The only common observed threat display toward humans is head toss-ing. The buffalo executes it standing in the alert posture, but sometimes may advance a few paces toward the threat. This behavior is displayed by adults and subadults of either sex, isolated or in groups. In groups, one or two high-status individuals, or a larger number of the animals, may display it when threatened. Most of the head tossers are in the front rank, but a few may be deeper inside the herd, stretching their bodies upward, either to gain a better view of the threat or to be more promi-nent while head tossing. Their posture is like one signifying intended flight in other circumstances. This may have relevance, since threat displays, when directed at humans, are normally followed by flight. Head tossing may be repeated by the same individual for at least 10 minutes, at irregular intervals of the order of 1 minute. Out of 63 records of this behavior, 41 (65 percent) were females. Head tossing was ob-served in buffalo groups alarmed by distant human sounds, with the threat out of sight. Buffaloes were observed to respond by head tossing to humans sighted, without the benefit of smell, at distances of up to 100 meters. If both sight and smell are used in identifying a human, head tossing may start from much farther away. No threat displays prior to open aggression against humans were observed.

Occurrence of aggression

The frequency of buffalo aggression against humans is difficult to mea-sure or discuss comparatively. An increase in the human population in a given buffalo area is normally accompanied by an increase of human predation on buffaloes, which will increase the chances of aggression. At the same time, the greater number of humans of itself increases the statistical chance of buffalo aggression against man. In the case of protected areas where game viewing takes place, there may be a large number of visiting humans, statistically speaking, but the interaction between them and buffaloes (or other wildlife) is of a special kind, which seldom leads to aggression. Buffaloes inhabiting areas where human predation is scarce are less aggressive with respect to man, irrespective of the level of their aggressive behavior toward, for example, lions.

Thus buffaloes recognize and learn to assess human threat. The next section describes observed buffalo responses to man-made sounds, which tend to bear this out.

Responses to human sounds

In the Chunga and Busanga areas of western Zambia during 1977 and early 1978, I allowed recently hunted buffalo herds to approach me downwind to within some 15 meters. At that point, I shouted intermittently for 30 seconds, in a variety of ways but without the use of lips. I did this on five occasions. In all cases the buffaloes halted and listened, then either remained stationary or continued to advance in my direction. I then spoke several words in a normal conversational tone; in all cases the buffaloes stampeded away.

In Kenya and Zambia, when I fired single rifle shots in the air near buffalo herds in localities where killing with firearms occurred, including a portion of the Kafue National Park where staff rations were shot, although the buffaloes were otherwise officially protected, an immediate stampede invariably occurred. However, on two occasions when a shot was fired near buffaloes deep within a protected area, in the Kafue National Park, they reacted both times by showing little agitation: The nearest animals ran away from the shot for some 20 meters or less, some of them halted, facing back to observe, then all resumed their previous activity, which in both cases was forward movement accompanied by low to moderate grazing. When the same herds moved into areas where killing with firearms occurred, they were never observed to behave in this way, but fled instead. The reaction to a gunshot may therefore depend on the expectation of human aggression.

On four occasions I observed buffalo herds overflown at medium altitudes by military jet aircraft. On one occasion no response was observed; on another, an adult female assumed the alert stance; on the remaining two occasions, both involving the same herd, the top male assumed the alert stance, facing away from the aircraft. A sonic boom, however, brought an entire resting herd to their feet, at the alert (one observation). Low-flying single-engine propeller aircraft most often panic buffaloes.

The ordinary range of thunderclaps does not stimulate responses like those brought on by the sounds discussed above.

Aggression without obvious provocation

Buffaloes occasionally respond to human intrusion by open aggression. I experienced such aggression in the Busanga area, Zambia in 1974. A high-status male was trailing some 100 meters behind his herd, at dusk. I stood at the base of a termite mound which the buffalo passed on the opposite side. I had caused slight noise by disturbing shrubs. Some 30 meters past the mound (as determined later by spooring) the buffalo turned around, walked quietly between scattered bushes directly toward me, and attacked when 7 to 8 meters away. I avoided the buffalo by making use of a tree that grew on the mound. The animal halted before the tree, then, after about 20 to 25 seconds, he turned abruptly in the direction of the herd, which was out of sight, and walked away. He emitted short low-pitched grunts, twice while charging and once as he moved to depart. I witnessed similar aggressive behavior by an adult male directed at Game Guard Stephen Chola (National Parks and Wildlife Service), as the latter was collecting firewood at dusk, east of the Ntemwa–Kebumba confluence, Kafue National Park, Zambia (1974). In both cases, sound appeared to draw the buffaloes' attention, while fading daylight and intervening shrubbery prevented them from seeing the cause of the sounds. This aggressive pattern resembles one that was observed in buffalo males making preventive attacks on trailing lions. However, in these particular circumstances, although air movements were slight, the aggressive buffaloes should have been able to smell the humans. Thus the above behavior may be a form of preemptive aggression, in defense of the herd, employed against any intruder.

Reciprocal aggression

There is no doubt that wounded buffaloes can be violently aggressive toward man. This is one of the most described behaviors of S. c. caffer, and a few of the many references on the subject inspire credibility (e.g., Selous, 1881; Uganda Game Ann. Repts. 1925–60). I observed this behavior, directed at me and others, in the Lolgorien, coastal hinterland, and Loita areas of Kenya between 1964 and 1967. In all cases but one, the wounded animal was hidden in dense vegetation and attacked when a human approached to within some 4 meters. In these cases the buffaloes appeared to locate the approaching human by hearing and sight. The buffaloes did not move below the neck, even when approached from the rear, until the human was within attacking range. In one case, a wounded mature male, hidden in dense shrubbery charged across some 12 meters of short open grass, while the human recipient of the charge was walking not toward the buffalo but across his field of vision. All the buffaloes observed while attacking depended strongly on their eyes to guide them and kept them on the recipient of the attack. If the recipient changed position, the attacking buffalo at once adjusted its charge accordingly. The top of the head was held slightly lower than the withers, muzzle well forward, until approximately 1 meter from the recipient, when it was lowered, to allow a hooking and upward movement with one horn or an upward toss with the head held approximately level. In successful charges the buffalo repeatedly horned the recipient, with apparent intent to kill, until shot or scared away. Despite the tenacity with which a buffalo often presses its charge, I have twice witnessed a previously wounded bull being scared away from an already prostrate victim it had charged by a nearby gunshot. In one case, the shot was fired in the air as both the buffalo and the man pinned down by the buffalo's head were invisible in dense shrubbery, for fear of hitting the man; in the other case, a small-caliber rifle was unskillfully fired in the direction of the buffalo goring the human recipient of a charge. I have, however, observed at least two other instances in which wounding gunshots fired at a charging buffalo failed to stop the attack until the animal was killed.

This type of aggression may be species-selective and not an indiscriminate response to any approaching large animal. I have observed a wounded but not immobilized adult buffalo male, which was repeatedly approached to within 2 meters by grazing roan antelopes (*Hippotragus equinus*) and hartebeests (*Alcelaphus lichtensteini*), in Zambia in 1977. The buffalo showed initial interest in the antelopes but there was no sign of aggressive behavior.

Aggression by females with calves

I have not witnessed open aggressive behavior toward humans by females in defense of their calves. However, my observations of behavior that I interpret as marginal to open aggression suggest that it occurs, and there are published references to such behavior (e.g., Uganda Game Dept. Ann. Rept., 1953; Sinclair, 1977).

Aggression by nonadults

Even small calves show aggressive behavior when threatened, but after having first shown flight. The attack is brief but executed with all the

Figure 8.1. An abandoned young male calf hiding in a bush at my approach. A moment later, he charged me violently. (Zambia, 1973.)

force available. The head postures are as in a charging adult. Unlike adults, however, calves do not persist in the attack. Two cases are described: A few-weeks-old male calf, left behind by the herd, tried to flee at my approach, then hid under a shrub (Figure 8.1). When I reached the hiding place, the calf charged me violently (Zambia, April 1973).

On another occasion, also in Zambia, in July 1977 when accompanied by R. A. Conant (then Wildlife Biologist, National Parks and Wildlife Service), I saw an approximately 2-month-old male calf that was trailing a herd in loose association with some adult males. The calf lagged behind and I approached it to within some 10 meters. It stood in the open, watching us, then turned in the direction of the herd. I continued to approach, and at about 2 meters the calf retreated a step, stopped with front legs spread wider than usual, then charged me with violence. I moved behind a sapling, and the calf went on its way at a walk–run.

Aggression against vehicles

I experienced aggression by an adult male while in a moving Land-Rover. I was driving very slowly, accompanied by an assistant, along a motor track pitted by elephant feet, near the Lushimba boundary post of the Kafue National Park in June 1977. An adult male buffalo which had been lying down invisible in high grass about 50 meters away stood up and ran toward the vehicle, adjusting his direction as the vehicle moved. When the buffalo was extremely close and no doubt of his intention could remain, I accelerated despite the elephant foot pits, leaving him behind. The buffalo turned into the motor track and followed (i.e., apparently chased) the vehicle for another 110 meters, then, outdistanced, left the track still at a run and halted about 75 meters from it, facing the Land-Rover. He took three steps toward it, halted, turned, and fled at a lope. I did not detect through binoculars any trace of wounding. The Land-Rover was of the color known as "Sahara yellow," the time about 1130 hours, and the light good. The herd to which this animal belonged was not in sight.

Five cases of buffalo aggression against vehicles are described by the Uganda Game Department (Annu. Repts. 1934–37). These reports generally inspire confidence, but unfortunately there are no comments on the probabilities of the buffaloes in question having been previously wounded, an important behavior-modifying factor.

Buffalo–lion encounters

Buffalo as prey

Lions are the only important nonhuman predators on the buffalo. The lion's preference for the buffalo as prey varies with the locality. Prey selection depends on a balance of local factors, such as the availability of prey species, their vulnerability, food yield per kill, and physiographic features which will determine animal mobility, distribution, and hunting or predator-evasion tactics. As prey, the buffalo has several desirable characteristics. It may occur in large numbers, is not migratory, moves through its home range in largely repetitive patterns, is easily outrun, and yields much meat per kill. At the same time, it is relatively dangerous and difficult to subdue individually and has developed aggressive-defensive social behaviors toward predators. On balance, though, the chances are good that locally the buffalo will rate high as prey, especially where lions can increase the safety factor by hunting in prides.

In the Lolgorien area and the coastal hinterland of Kenya during 1964–67, where lion populations were fairly small and shifting, all buffalo kills that I noted suggested accidental predator–prey encounters. The Lolgorien lions hunted mainly zebra (*Equus burchelli*), topi (*Damaliscus korrigum*), and kongoni (*Alcelaphus b. cokii*); less often waterbuck (*Kobus ellipsiprimnus* and *K. defassa*, both of which occur in the area), and buffalo. In the coastal hinterland region, kongoni, zebra, and fringe-eared oryx (*Oryx beisa callotis*) were the most common prey, with eland (*Taurotragus oryx*) and buffalo also recorded. Schaller (1972)

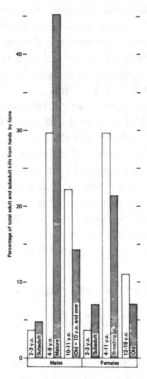

Figure 8.2 Lion predation on buffalo herds. Comparison of adult and subadult kills from herds in the Serengeti area, Tanzania, □ (based on data derived from Sinclair, 1977), and in the Busanga area, Zambia, ■. Infants and juveniles were excluded because means of comparison were doubtful. However, these categories were also present among kills in both localities.

guesses that in the Serengeti area of Tanzania buffaloes represent only 15 percent of the lion's food. In the nearby Manyara National Park, however, he reports that buffaloes represented 62 percent of all the 1967–69 lion kills, possibly because buffalo constitutes the major available prey. In Zambia, buffaloes tend to be the main lion prey. In the Mkushi district (1974) and probably in the general area of the Kafue National Park, but certainly in the part of it lying north of the boundary between the Southern and Central Provinces, and in the neighboring districts, during 1970–78, lions were positively selective toward buffaloes as prey. In the portion of the Kafue National Park lying within the Southern Province, there may be lions that hunt the hartebeest (*Alcelaphus lichtensteini*) preferentially to the buffalo. Mitchell et al. (1965) report that 30.5 percent of lion kills in the Kafue National Park during 1960–63 were buffaloes, followed by hartebeests (16.3 percent). In the northern sector of the park in 1970–78, I estimate that buffaloes were a substantially greater proportion of lion's prey than 30 percent.

Table 8.1. *Buffaloes killed by lions in the Busanga area, western Zambia, according to sex and age (herd animals only), 1977*

Sex and age group	Number	Percentage	Total according to sex
Male calves/juveniles	4	8.2	
Male subadults	2	4.1	
Mature vigorous males	19	38.8	
Old males	6	12.2	
			All males: 31 = 63.3%
Female calves/juveniles	3	6.1	
Female subadults	3	6.1	
Breeding females	9	18.4	
Old females	3	6.1	
			All females: 18 = 36.7%
Total sample	49	100.0	

In the Zambezi Valley in recent years, the preferred prey species appeared to be the buffalo and kudu (*Tragelaphus strepsiceros*). In the Mabalauta area of Zimbabwe in 1972, kudu, waterbuck (*K. ellipsiprimnus*), and buffalo were predated, with the last clearly an important, if not the main, prey. In South Africa's Kruger Park buffaloes accounted for only 3.9 percent of the lion kills recorded during 1954–66, but kudu, as in Mabalauta, was a fairly important prey (Pienaar, 1969b).

I tabulated 49 lion kills of buffaloes from the Busanga area of western Zambia, according to their sex and estimated age (Figure 8.2). All the kills were herd animals, that is, any individuals living apart were excluded, although stragglers moving together with herds were included (Table 8.1). Sinclair's (1977) breakdown of 75 Serengeti kills according to sex and age was compared to these Busanga herd kills. Males 12 or more years of age, which he states live apart from herds in the Serengeti, were subtracted, reducing the sample size from 75 to 54. In this adaptation of Sinclair's data, I classified his "age 10–11" males with my "old"[1] and his "age 10–11" females with my "breeding" (Table 8.2). My Busanga data, abbreviated to percentage totals of males and females, including calves and juveniles and excluding them, were compared with figures by other observers, for various African localities (Table 8.3). The figures in Table 8.3 are not strictly comparable, but they all indicate a preponderance of males among the kills. The male/female ratios are not comparable, without making exclusions, because some samples represent any buffalo kills and some represent only herd kills. The statisitical similarity of these figures suggests an underlying bio-

[1] Including 10-year-olds and up for a better fit with Sinclair's age subdivisions.

Table 8.2. *Sex and age distribution in lion kills of S.* caffer *in Serengeti and Busanga herds*

Sex	Age group Sinclair[a] (yr)	Mloszewski	Serengeti[a] Number	%	Busanga[b] Number	%
Male	2–3	(subadult)	2	3.7	2	4.8
Male	4–9	(mature)	16	29.6	19	38.8
Male	10–11[c]	(old 10+)	12	22.2	6	14.3
Female	2–3	(subadult)	2	3.7	3	7.1
Female	4–11	(breeding)	16	29.6	9	21.4
Female	12–19	(old)	6	11.1	3	7.1
			54	99.9	42	99.9

[a] Adapted from Sinclair (1977).
[b] Present study.
[c] All age 12 and older males excluded.

Table 8.3. *Percentages of S.* caffer *males and females in lion kills in national parks in South, Central, and East Africa*

Locality	Comments	Number	Percentage males	Percentage females
Busanga area, Kafue, Zambia, 1977 (present study)	Including calves and juveniles	49	63.3	36.7
Same as above	Excluding calves and juveniles	42	64.3	35.7
Kafue, Zambia 1960– 63 (Mitchell et al., 1965)		92	60.9	39.1
Kruger, South Africa 1954–66 (Pienaar, 1969)		286	67.5	32.5
Manyara, Tanzania 1967–69 (Schaller, 1972)	Large animals	50	88.0	12.0
Serengeti, Tanzania (Sinclair, 1977)	Excluding calves	75	68.0	32.0
Same as above	Excluding calves and males age 12 or more	54	55.6	44.4

logical significance. However, it appears that the similarity, if not entirely accidental, is due to deep-seated complex factors and not to a simple relationship common to them all, because some are the product of one type of buffalo–lion interaction, whereas others are the product of a different type of such interaction, depending on whether predation is buffalo-selective or not.

The lions' hunting patterns depend on the local importance of the buffalo as prey. When the buffalo rates low on the lion's prey list, lion–buffalo encounters are largely fortuitous, and the buffaloes killed are the ones most easily encountered accidentally, that is, scattered bachelor males and strays around herds. The strays include a larger proportion of males than statistically expected from herd structures, where I observed male/female ratios (all ages) ranging between 1.00:1.17 and 1.00:1.55, because males tend to be peripheral or separate from the main body of the herd more often than females. Thus in predation dependent on chance encounter males may be expected to predominate among kills, as actually observed.

In localities where the buffalo is the main prey, lions develop a hunting routine that brings about lion–buffalo encounters not by chance but by design. The predation is not dependent on chance encounter, that is, multispecific, but rather species-selective for buffalo. That is, the lions do not prowl about until they meet any potential prey, which may sometimes chance to be a buffalo, but seek buffalo herds. In this kind of interaction, a buffalo herd's tendency to move along established routes is of importance, with advantages to both predator and prey. The lions can lie in wait along the buffaloes' route or intercept the herd, and they are able to select localities on its circuit where stalking is easy, because of good concealment, poor wind conditions, or a propensity of the buffaloes to behave in some way that favors the hunt. The lions are able to kill at need, with a minimum of search, that is, with a relatively low energy outlay. On the other hand, the buffaloes' knowledge of the route allows them to implement protective measures where lion attacks are likely to occur. Where predation on the buffalo is heavy and lions systematically stalk the same herds, the protection of breeding females and young becomes important. Because the attacks are expected, these weaker but essential herd members tend to be protected in the lions' hunting localities, while the least useful herd members, that is, mainly stray males and invalids, fall prey to lions. The result is a two-sided payoff: To the lions in lower energy outlay and meals on time, to the buffaloes in survival selection favoring the most essential and capable individuals.

I argued above that where predation is dependent on chance encounter, the preponderance of males among kills may be explained by the greater probability of lions meeting male buffaloes, and that when a herd in which females predominate is the lions' principal prey source, one must look for a different reason why in these circumstances also there is a preponderance of males among the kills. The question arises why there are not more immature individuals killed, as their size is more commensurate with lions' killing abilities. According to Schaller (1972), solitary lions in the Serengeti tend to prey on animals weighing 50 to 300 kilograms, that is, one-half to twice a lion's weight. Therefore an adult buffalo of the Cape variety, weighing in the range of some 425 to 870 kilograms, is beyond the limit of an easy kill, even in some cases of participation by more than one lion, and several minutes, if not longer, may be required to seriously disable such an animal. On these grounds alone, if allowed a free choice, lions that systematically prey on buffalo herds may be expected to kill many calves, juveniles, and subadults, as well as many females, which in herds are more numerous than males. Lions may also of course be expected to favor various strays on the periphery of the herd. Figures show, however, that here, just as in areas where lion predation is not buffalo-selective, more males than females are killed and the proportion of nonadults killed is not exceptionally high. Table 8.2 indicates that breeding females and all subadults killed in the Busanga herds amount to 33.3 percent of all herd kills, while in the Serengeti the same categories amount to 37.0 percent of all presumed herd kills. Therefore the proportion of breeding females and subadults killed in the Busanga herds, where lion predation is buffalo-selective, appears to be about the same as, or even slightly less, than, in the Serengeti, where lion predation is multispecific or other-species-selective. The reasons can be found in the highly developed antipredator behavior of herds, which deflects the lions' interest from nonadult and female herd members and toward strays. Whereas in cases of multispecific predation antipredator behavior does not appear to be essential for the preservation of nonadults and breeding females, because the lions' interest is held largely elsewhere, in cases of buffalo-selective predation antipredator behavior appears to be very important for population survival.

Lions that do not kill buffaloes most of the time are not particularly good at it. Thus I have observed that western Zambian lions are more efficient buffalo killers than Serengeti lions. Since the sizes of both predator and prey are similar in the two localities, the difference in the killing ability must be real. Lions that specialize in killing buffaloes become very proficient, not only in pulling down the prey but also in all the preliminaries leading to the kill. This proficiency appears to be learned and, given time, probably therefore subject to modifications under new environmental pressures. Nonetheless, some lions that for generations have preyed on buffaloes are so selective that they tend to bypass other, nearby, potential prey, for example, herds of zebra (*E. burchelli*), wildebeest (*C. taurinus*), roan (*Hippotragus equinus*), or lechwe (*Kobus leche*), while stalking more remote buffaloes. This behavior is explicable in terms of a hypothesis that, given an overall major advantage in feeding mainly on buffaloes, it pays to cultivate the somewhat specialized buffalo-killing proficiency, at the price of occasional opportunistic easier kills of other species.

The best example of buffalo-selective predation known to me is the Busanga flats area and vicinity, situated mainly within the Kafue National Park, at its northern end. There, buffalo-selective predation was observed even when search for buffaloes was necessary, while other species could be stalked directly. There were a few exceptions when buffaloes were absent and another prey species easily available. An exception is described in Chapter 7. I also observed young lions in the Busanga chasing and killing warthogs (*Phacochoerus aethiopicus*), lechwe (*K. leche*), and civets (*Viverra civetta*); the last were not eaten, possibly because of their musk. I also found a puku (*K. vardoni*) apparently killed by lions but uneaten for reasons unknown. Lion kills and attacks on buffaloes in the Busanga area were observed at different times of day and night, for example, 0200, 1100, 1600, 1900, 2000, 2300 hours. In 1977, the main resident pride, preying largely on the central Busanga herd, included 16 adults and subadults, and 1 juvenile, plus up to 3 cubs observed at different times. In all its hunts, this pride was led by the same two lionesses, while the largest male was usually one of the last, although he occasionally did participate in killing. One of the two lionesses was always well forward during a stalk or first to rush from an ambush. Several others joined her soon after she caught a buffalo. Some of them bit and held the victim by the hind legs and tail, one – often the leading lioness – stayed on the withers, pressing down on the neck, and nearly always one held the muzzle in the jaws, with a resultant diagnostic perforation or chip in the premaxilla or maxilla in the majority of lion kill buffalo skulls from that area. Simultaneously, pressure was put on the buffalo until it fell. This was the method used on adult buffaloes, on the margins of or behind the herd. Lions that ventured near the main body of a herd risked attack by one or more buffaloes, unless the herd was running. Most nonadult kills were either strays or made in a stampeding herd, when the lions, independently or in pairs, penetrated it, lashing out at small animals or straddling them, until either the buffalo or the lion fell down. This behavior was difficult to observe in the dry season because most of the action was hidden in clouds of dust. Often the lions at the rear of the running herd took possession of downed nonadult buffaloes before the killers returned. A lion ambush often resulted in a stampede. I will describe three kills made by this pride.

On the morning of August 21, 1977 I was accompanying the central Busanga composite herd, about 700 strong at the time, as it moved northward across grassland toward the Kasolo Kampinga thicket, an ecologically very important 3 × 1 kilometer relatively high area on the floodplain, which was one of the lions' favorite killing places. Normally, with antipredator precautions, the herd penetrated grassy embayments in the thicket, and even the thicket proper, to browse and rest. This time, however, it bypassed the thicket at its southeast end, with much alert behavior, apparently responding to fresh lion scent. The herd then swung east-southeast (0906 hours) and walked toward a pool west of the Lufupa main channel, some 3 kilometers from the thicket. I scanned for lions from the top of a termite mound and saw several move rapidly east, away from the thicket, swing southerly, and follow approximately the Lufupa drainage, converging on the pool toward which the herd was walking. The herd, following a usual route, arrived at the pool at 1038 hours, in a column, and partly encircled the pool, to drink. The lions had reached the location previously and hidden in a large mound on the downwind side. A stray male buffalo caught their scent, turned diagonally toward the herd and fled, causing the rear portion of the composite herd, about 300 buffaloes, to stampede back in the direction of the thicket, with the two lionesses, followed by others, in pursuit. When the running buffaloes were some 200 meters from the pool and the buffaloes still standing near it, several lions pulled down a stray male, while others penetrated the running herd from the rear and one swiped a calf to the ground (1048 hours). The other 300 or 400 buffaloes that remained standing by the pool kept at the alert until about 1100 hours, then drank and continued their easterly walk. The buffaloes that stampeded slowed down to a walk by 1051 hours, made a wide turn, and also started moving easterly again, but remained separate.

By August 25, 1977 the two herds had again converged and were together in the open, partly swampy grassland of the Shindende location, east of the main Lufupa drainage and some 13 kilometers east of the Kasolo Kampinga thicket. At 1500 hours the buffalo herd was resting in the open, with many lookouts, and about 8 wildebeests at the northern edge of the herd. Some 100 meters north of the buffaloes grazed a further 32 wildebeests, and over the 75 meters extending directly north of them were scattered 27 zebras. Grazing back and forth over much of the area were some 18 roan antelopes. Just west of the antelopes and zebras was one of the Lafupa drainage troughs, about 1 meter deep, aligned about southeast and passing near the edge of the buffalo herd, and nearly dry except for one fairly large water remnant. At 1600 hours I saw the large lion pride leave a mound some 1 kilometer away and pad southeasterly toward the grazing and resting ungulates. The two hunting lionesses were in the lead. By 1622 hours, the lions were walking inside the drainage trough, only their backs showing, and passed within some 17 meters of the nearest zebras, which noticed the lions but displayed no flight or prolonged alert, and continued grazing (1625 hours). When the lions were some 60 meters from the buffaloes, which were now stirring and spreading after resting, the lead lioness crawled out of the trough, two or three times her own length toward the buffaloes, while the rest of the pride remained passive in the trough (1633 hours). The leading lioness was watching closely and the second hunting lioness also crawled out of the trough, watching. At 1642 hours, a buffalo gave a vocal warning, occasionally used in big herds when threatened by lions, a uniquely extended relatively high-pitched "waaaaaa." I had felt a gentle air turbulence a moment before, which was probably responsible for the detection. The buffaloes immediately compacted, a large group of adults of both sexes at the alert facing the lions at first, and the herd started walking away rapidly, in a compact body. One lioness followed for a short distance while the other lions remained passive. At 1708 hours, the buffaloes were moving east, water-vocalizing (Chapter 11) while the lions rested around the water remnant in the drainage trough. The grazing zebras and antelopes, which otherwise may have drunk at that spot, started drifting away toward another wet area, and that, aside from the occasional routine alert stances, was the only antipredator behavior on their part that I noticed. At 1711 hours, a leading lioness moved in the wake of the buffalo herd, followed closely by the others. At 1729 hours, the herd's rear was beginning to spread out, and a single adult male was walking considerably to the left of the herd. The lionesses rushed him, instantly followed by the others. The buffalo started flight but was immediately blocked and brought down in less than 2 minutes.

The elaborate but largely effortless preparation, followed by a rapid kill, was frequent. The ability of western Zambian lions to rapidly disable buffaloes explains multiple kills, such as described by Ansell (1969) as having occurred in the south of the Kafue National Park, where six adult buffaloes were killed by a pride of lions, in undulating grassland, apparently at an average distance from each other of not more than 300 meters and very likely less. The male/female ratio in that kill was 2:1.

However, inefficient killing was also observed, although exceptionally. On June 25, 1977, at the northeast edge of the Kasolo Kampinga thicket, at 0123 to 0149 hours, I was present at a fight between a very large female buffalo, whose horn spread I later measured as over 132 centimeters, and lions. Although I was very close, in a tree, night visibility was poor. The buffalo herd, which had been grazing and watering almost at my feet, apparently got wind of the lions as they approached and stampeded a long distance away from the thicket, where, to judge by sounds and later spooring, it formed a compact defensive group. As the herd stampeded, the younger lions chased it, and apparently remained watching it when it stopped. The old buffalo cow, however, made a stand, and two adult lions attacked her. The lions' growls and the cow's grunts accompanied a great deal of noise caused by various kinds of impact. Finally, with the return of other lions, the buffalo was pulled down, after at least 26 minutes of struggle. In the morning I noted that two lions were injured, one in the head and ear, the other in the hind quarters, though not critically.

There were other lions, besides this large pride, in the same general area during 1977, whose movements overlapped with this pride's, but even those that hunted in pairs only were observed to trail buffalo herds. The only exceptions that I have noted were an old male and female, frequently seen between Moshi and the Big Tree camp, south of the Busanga, which gave the impression of preying mainly on puku.

Antipredator behavior

Antipredator behavior changes from one buffalo herd to another, since it is largely an adaptation to local conditions, but it can all be classified in one of six basic categories: increased caution, selective scrutiny, auditory-olfactory probing, flight, preventive aggression, and reciprocal aggression. In buffaloes that are not hunted systematically, only flight and reciprocal aggression are common high-intensity antipredator behaviors.

I use "caution" to describe the wide range of low-intensity responses to anticipated but unconfirmed threats. The most common displays of caution are a compact formation, with the number of strays sometimes reduced to nil, and the more or less continuous recurrence of the alert stance in the herd, even when walking. A buffalo standing or walking at the alert holds its muzzle relatively high, a posture that may be related to the quality of a buffalo's middle-distance and long-distance vision, thus having a direct functional cause (Figure 8.3 a and b). Caution may take the form of bypassing a possibly dangerous part of the habitual route, as already illustrated earlier in this chapter, leaving a locality rapidly or dallying in one.

Selective scrutiny is a mainly visual response, to localized unexplained motion of unknown threat potential, or in anticipation of a threat from a particualr location. It is a more specific behavior than caution, always done while stationary, the head angle varying with the distance and slope of ground between the scrutinizer and the point of interest. The scrutinizer's body may be either in line with or at an angle to the line of sight (Figure 8.4). Sixty percent of observed scrutinizing behavior was displayed by females ($n = 180$), that is, a male/female ratio of 1.0:1.5, which falls within the range of observed male/female ratios in herds (1.00:1.17 to 1.00:1.55). In herds moving through localities where lion predation is frequent, single adult animals halt, turn the head or the whole body to face rearward, and scrutinize the ground

a

b

Figure 8.3 The alert or scanning stance in adult males: (a) profile and (b) diagonal views.

a

b

Figure 8.4. (a) A male in a "flight or fight" stance. (b) A male displaying mild curiosity.

behind the herd for periods from 6 to 80 seconds. The behavior is adapted to the topography, because, when scrutinizing to the rear from a ridge or hilltop, the animal also takes a moment to look forward, over the heads of the other herd members. In herds crossing undulating or hilly ground, 62 estimates were made of the elevation, relative to all nearby terrain in sight, of positions from which scrutinizing occurred. The results suggest that, where the terrain is uneven, high spots are preferred for scrutinizing (Table 8.4).

Individuals were observed to halt and face backward while in the middle of a moving herd, remain in this position and scrutinize rearward as the herd moved past the spot where they were standing, then turn around and follow. A scrutinizer that has sighted a distant threat assumes the alert standing posture, often takes a series of quick short

Table 8.4. *Scrutinizing behavior and utilization of topography by S.* caffer
(n = 62)

Estimated relative elevation of visible part of herd's route	Observations	
	Number	Percentage
Topographically lowest 25%	8	12.9
Topographically intermediate 50%	26	41.9
Topographically highest 25%	28	45.2
Topographically highest 5%	7	11.3
Topographically lowest 5%	1	1.6

steps forward, maintains the alert posture for a moment afterward, then swings around and runs in the direction of herd movement. This generates a stampede. If the locality is a lion's preferred hunting ground and familiar to the buffaloes, the stampede may continue until the momentum is lost through tiredness of the small nonadults. On other occasions, possibly when the meaning of the threat message is not clear, the stampede is stopped after only a short distance, and immediately several adult males, usually with some females, work their way to the rear in an alert-defensive move, while some, but not necessarily all, other herd members also turn to face the rear. If the threat is confirmed, the normal sequel is a resumption of flight, in a direction imposed by some of the pathfinders. During halts, scrutinizing is done in a recumbent or standing posture, and the response to a sighted threat is similar to that described above. The alert standing posture and the quick few paces advance bring about alertness in other individuals, and wheeling as if on the run by the scrutinizer causes all herd members to stand up. If the scrutinizer continues to run, there is immediate escalating movement in the herd that results in flight. If the scrutinizer wheels as if on the run but immediately stops and turns again toward the source of alarm, the herd stands at the alert until further signals are passed.

I limit "probing" to the examination of the surroundings mainly by auditory-olfactory means. It is a distinct behavior in *S. c. caffer*. It may be observed when a pathfinder probes before leading a herd over relatively unfamiliar ground. The animal stands well in advance of the herd, a position compatible with decreased modification of air movement and auditory obstruction from the mass of the herd, and remains motionless for many seconds, sometimes extending into minutes. The behavior has been observed to last up to 8 minutes, although often it takes much less time. At the end of it, the animal usually steps forward and the herd follows. Occasionally the pathfinder elects not to move in the previously intended direction, either by remaining stationary until the herd spreads out to graze or rest, or by leading the herd in a different direction. It is assumed that this happens when a threat is sensed in the direction in which movement was originally intended, and there is much evidence that this assumption is correct. Probing is done by standing individuals during halts and may then involve an occasional change in direction of subsequent movement. However, it was seldom seen during halts unless the herd was under some extra-specific stress. Probing is common among isolated individuals, for example, bachelor males.

Flight is the most common initial response to intense lion threat (Figure 8.5). It is probable that in general, when prospects for success-ful flight are good, it is the prevalent response to intense extraspecific threat. I interpret buffalo aggression against another species as a way of improving the odds for successful flight, where the limiting cases are death, making further flight impossible, and threat elimination, mak-ing further flight unnecessary. All observed cases can be interpreted within the following basic sequence, which may be repeated or have stages omitted, but without a change of order:

Stimulus 1		Reponse 1		Stimulus 2		Response 2	
extraspecific threat	→	intended flight	→	intensified extraspecific threat	→	flight	→

Stimulus 3		Response 3	
interference with flight	→	defensive aggression	→ . . .

This sequence eventually terminates with either the death or survival of buffaloes; if survival, then either with or without serious injuries; if with serious injuries, then the survival potential is lowered. The only certain way of avoiding death or serious injuries is successful flight, which, using the games theory concept, is the highest possible payoff in this dual game. The type of behavior described in the earlier section "Aggression without serious provocation" is anomalous in this context if the threat is taken as directed against an *individual*: flight and not aggression would be expected. But if the threat is taken as directed against the herd, of which the aggressive animal is a vigilant compo-nent, then the behavior fits the above pattern, because the *herd* may not be able to flee successfully, and thus the behavior is defensive aggression in response to stimulus 3. Compatibly with this, only indi-viduals trailing or flanking herds at some distance were observed to show this type of aggressive behavior.

In the herds of the Mabalauta area, Zimbabwe in 1972, the threshold for the flight response appeared to be about the same for lion and human threat, although this is often not the case. If one of these herds became alarmed while on the move and the danger was not clearly

Figure 8.5 Buffaloes running at full speed, as they flee from lions.

identified, a stampede could result, but it did not always involve the entire herd, only the animals on the threatened side. If the threat lay ahead of a moving herd, the leading animals either halted, scrutinized for a few seconds, then turned and ran off, or, more often, wheeled around and ran off instantly. The individuals immediately behind the leaders imitated this action. Throughout this activity, the rear portion of the herd was still advancing in the original direction, but halted on realizing that the forward animals were returning toward them. After retreating approximatlely 50 meters from their most forward position, the leading animals halted, readvanced, and proceeded to examine the reality of the suspected threat, while the rest of the herd stood still, heads up. This phase usually lasted only a few seconds, at the end of which the leading animals, followed by the entire herd, sometimes resumed flight, probably in a divergent direction. If the scrutiny lasted longer than a minute, it usually signified that the herd would continue forward. In such cases, the lapse of time before the herd resumed movement could amount to many minutes. If the herd was stationary at the time of first suspecting danger, all the individuals usually retreated at a run before any halt was made to further scrutinize the suspected threat. If the threat was accepted as real, either immediately or at the end of a confirmatory scrutiny, the herds ran for up to a kilometer, then halted and usually started to walk again in the general prealarm direction, but by a roundabout route. During three weeks of observations, I recorded only one instance of a herd being diverted for long from its normal path by flight from a threat. The herd was being stalked by lions (October 9, 1972).

A strategy of prolonged flight succeeds sometimes, because lions may turn to another buffalo herd if prevented from making a kill for a sufficiently long time, or, if a chance opportunity arises during the pursuit, may make a kill from some other species. The prolonged flight strategy, however, is tiring for calves and juveniles, and can be employed only occasionally, since gain from avoiding a kill may be more than counterbalanced by loss caused by exhaustion of calves and juveniles. I will illustrate antipredator behavior involving long intermittent flight with an occurrence that I followed, by direct observation and careful spooring, on July 26–27, 1974, near the western perimeter of the Busanga area, Zambia.

The Lushimba herd, over 200 strong at the time, scented approaching lions while resting before dawn on July 26. The herd interrupted its rest and ran for 1.3 kilometers, then walked rapidly for 3 kilometers, changing direction several times. During the following daylight hours, the lions did not harass the herd, which, however, displayed more than usual alert and probing behavior, rested for a shorter time than usual (45 minutes in midmorning and 1 hour 58 minutes in the afternoon, i.e., 2 hours 43 minutes during the entire period of daylight), but grazed adequately. The herd watered at a large waterhole, beginning at 1826 hours, then grazed slightly and started to settle for a rest. Lions were heard from 1900 hours onward. Their first vocalizations were from far away but approached gradually, until by 1941 hours they sounded close. The herd interrupted its rest at the waterhole and walked rapidly away. The lions followed. Throughout the night, the herd moved rapidly, with few halts, little grazing, and two occurrences of running from the lions, one of about a kilometer, the other 600 meters. By 0750 hours on July 27, the herd went to rest, with some evidence of exhaustion, but the lions had left. The buffaloes went to rest on a slope, in a small patch of woodland, with several hundred meters of open grassland immediately upslope, and also upwind, from them, and a large open space downslope, and downwind. The entire herd was so well concealed and quiet within the small woodland patch that, although I knew they were there, I had difficulty discerning them. There was a large number of buffaloes watching all the time. An ideally situated termite mound made possible a very close undetected approach. The gap between the buffaloes and the mound, some 12 meters, was intensely watched by never fewer than two adult males. This mound was the only feature partly obstructing a wide view. I took photographs to record the utilization of woodland for concealment (Figure 6.14). By 0926 hours, slight movement and rare quiet vocalization began within the herd. By 0937 hours, a female and two juveniles were at the edge of the upslope grassland, and by 0942 hours a few juveniles were just out of the woodland, grazing. From then on, vilgilance dropped to an ordinary level.

In order to observe the reaction to threat, I requested Game Guard S. Chola, who accompanied me on this occasion, to make a wide detour, stalk the herd from the upslope direction, and then show himself. He was detected while invisible to the buffaloes, at some 150 meters, with the help of a weak diagonal wind: three adult females and one juvenile stood at a rigid alert facing in his direction and scrutinizing (simultaneously, however, a subadult male walked out casually into the open, to graze). When Chola showed himself, at 80 to 90 meters, quietly, all buffaloes stood up within 4 seconds, "zeroed in" on him in about 5 more seconds, moved away from him and past my mound, halted for 5 to 6 seconds, ran for about 110 meters, then walked away rapidly.

I am unable to generalize, from my observations, on the buffalo–lion flight (fright) distance, that is, the distance at which, given only the threat of proximity, avoidance behavior starts. This appears to vary both between and within regions. In the open country east of the Mara River, Kenya, I observed a buffalo herd move away from a small lion pride that included two prominent males, when passed within 24 meters between the closest lion and buffalo positions, which I later measured. In other cases, this distance varied between some 40 to 100 meters. Schaller (1972) states that in the Sereneti lions may pass within 20 or 30 meters of watching buffaloes.

Protective aggression, consisting of similar behavior to that described in the section on unprovoked aggression against humans, was also directed against lions. On September 8, 1977 at about 1620 hours, I was in the middle of a dry *dambo* (seasonally wet untreed grassland drainage area), just north of the Moshi headwaters in the Kafue National Park, Zambia, watching the main Ntemwa buffalo herd move along the *dambo's* edge, past several large bush-covered termite mounds, which were on the far side of the buffaloes from my position, and on their left with respect to their walking direction. The herd was trailed at a distance of some 100 meters by three large bulls, which were walking to the left of the herd's path, that is, approaching the distal side of the mounds. All three halted about 20 meters short of a mound, all at the alert, one displaying intended flight by taking a jerky step to the side and adopting an intensely alert posture, head high, outstretched and twisted slightly to the side. Seconds later, one of the other bulls advanced quietly toward the mound, then accelerated to a run, grunting. A few seconds later he disappeared from my view behind the mound, and an adult male lion came into view on the other side of it, loping, the bull loping behind him for another 50 meters. Then the aggressive buffalo walked at a fast pace toward the herd, the other two bulls, close together, doing the same.

On January 10, 1978 at about 0900 hours, between the old Chunga pontoon and the new bridge across the Kafue, on the eastern edge of the Kafue National Park, I followed a large buffalo herd through scrub-covered wet grassland, within sight of the river. The herd suddenly stampeded, in the same direction in which it had been walking. Then I noticed three bulls, now well behind the herd, running at almost right angles to the herd's direction of movement, toward adjacent slightly higher and drier ground. I scrambled up a tree just in time to see the three bulls send at least two young adult lions scuttling off. The buffaloes were so close to the lions that they were making horning motions and must have hit them. I followed, but by the time I reestablished contact, the buffalo bulls were on their way to rejoin the herd, and the lions out of sight. This pattern of aggression appears intended to discourage lions from trailing close behind buffalo herds.

When herds go to rest in localities where lion attack is probable, herd males sometimes conceal themselves motionless in dense cover, singly or in small groups, at some distance from the herd, and on a side from which lions could approach unnoticed. One morning in September 1977 in the Busanga area, I spent 2 hours 18 minutes on a branch, two meters above four such bulls, and noted that, although they were nearly motionless, no fewer than two at a time were intensely watchful throughout, neither ruminating nor appearing to relax. On this, as well as other similar occasions, the bulls remained in these attitudes for not less than 10 minutes after the resting herd moved away. They then stood up and moved without delay after their herd.

Reciprocal aggression, that is, a fighting response to direct attack, is frequent in buffaloes attacked by lions, and observations have been published in the past (e.g., Cullen, 1969; Schaller, 1972). These observations, with my own in addition, indicate that the majority of such aggressive confrontations ends in the buffalo's death, but that this is by no means always the case. The following three incidents, observed by the same reliable witness, in the same region (Luangwa Valley, Zambia), over a period of only 18 months, suggest that lion killings by buffaloes cannot be dismissed as merely isolated events (J. E. Hazam, personal communications).

One evening in February 1972 Hazam, while employed in the Luangwa Valley National Park, Zambia, was in a hide a few kilometers south of Mfuwe. He saw two old buffalo bulls, running, one of them whirling around repeatedly, with a male lion hanging onto his neck and shoulders, while a second lion followed. The bull tried to shake off the first lion by means of the whirling action, and succeeded, reaching the lion with one horn. The lion was thrown down and the bull, still whirling around, trampled it. In the meantime the second lion, which was trying to engage the whirling bull, was knocked down by impact with him. The other bull charged in and gored the second lion, twice. The two buffaloes trotted off, one of them visibly wounded. Both lions were gored in the belly, with much blood and flesh in evidence. One lion walked away immediately, though with great difficulty, while the second remained sitting, then walked away as well, halting frequently. Although the lions were not seen dead, the witness judged that the nature of their wounds made the lion's survival unlikely. On the following day, Hazam encountered the wounded buffalo, alone, in the same area.

In May 1973 one evening at dusk, near Chilangozi in the southeastern corner of the Luangwa Valley National Park, Hazam also saw a prostrate immobile male lion and the rear of a buffalo herd retreating at a run. The witness did not carry a firearm and thus did not leave his Land – Rover and walk up to the lion, but drove very close to him, with the headlights on. The lion had blood in the neck and shoulder region and appeared newly dead.

In August 1973 Hazam drove around a turn in the road overlooking grassland beside the Kapamba River and saw 4 lions starting an attack on a buffalo, one of a group of 25 bulls. One lioness was thrown in the air by the attacked buffalo, which was intermittently harassed by a second lion. The buffalo did not charge the lioness he had tossed, but the lions appeared to give up, while the buffaloes fled. The tossed lioness was dead, with a large swelling on her right side but no open wound.

Buffalo–elephant encounters

Threat behavior toward elephants

Agonistic interactions between buffaloes and elephants were observed to occur very often when the two species met. In most cases the aggressors were elephants. Only 10 percent of all agonistic interactions between buffaloes and elephants, for which the beginning has been observed, was initiated by buffaloes. Buffaloes initiated threat behavior toward elephants in the following observed instances.

The lateral threat posture was assumed by a mature male (one of four bachelors) toward an adult elephant approaching diagonally from upwind, in open acacia woodland, in the morning, in the Serengeti National Park, Tanzania, 1973. The elephant spread its ears a few times but appeared to pay no attention to the buffalo. The latter maintained the threat posture while the elephant reduced the distance from 17 to 6 meters then moved restlessly more or less in place, and with one brief head-tossing display walked off, after turning through an angle of about 160° away from the advancing elephant. A very similar pattern was observed one evening near water in the Tsavo West National Park, Kenya, 1965, which involved a mature buffalo male attached to a small herd. In Zambia, 1972, a female with a calf was observed to advance out of the herd, a total of eight steps, toward three elephants which just arrived to water some 20 meters away. She assumed the alert stance and went through three head-tossing displays, then retreated into the herd.

Thus, observed agonistic behavior toward elephants that is initiated by buffaloes consists of fairly low-intensity threat displays followed by retreat.

Response to elephant aggression

Elephant threat displays toward buffaloes are usually boisterous and suffice to make a group of buffaloes give ground. The buffaloes may pause, with some individuals at the alert, before retreating. Alert stances facing the elephants, with occasional head tossing, largely by high-status females, are frequent during this kind of retreat, but a few individuals may even graze at the same time (not a displacement activity).

Occasionally, however, elephants introduce high stress into the entire buffalo herd, causing outright flight, by displays that include much trumpeting, squealing, and charging at the buffaloes. A good example was observed east of the Ntemwa River in the northern sector of the Kafue National Park, Zambia, 1974. A buffalo herd of about 370 was resting by a waterhole in the afternoon when a cluster of seven elephants arrived, composed of three females, two juveniles, and two barely adult males. Some 20 meters from the buffaloes the two male elephants started a loud threat display, then charged at the buffaloes, which were by then clustered together at the alert. The buffaloes nearest the elephants began to flee around the herd's edge, while the rest of the herd turned tail and fled for about 100 meters. The two elephants did not quite catch up with the herd and returned toward the waterhole. However, noticing that the buffalo herd was still nearby, they repeated the display and the charge, this time shoving two of the buffaloes along from the rear. Apparently, the buffaloes tried to remain near the waterhole to drink but the two elephants induced them, by repeated charges (four in all), to depart after some 40 minutes of intermittent aggression. At each charge, the herd was stampeded into a run. This is the only time that I observed physical contact between buffalo and elephant in an agonistic encounter. However, several buffalo–elephant encounters ending in a buffalo's death have been reported by the various East African game and national parks departments (see Cullen, 1969).

Aggression directed at cattle egrets (Bubulcus ibis)

Buffalo aggression toward cattle egrets was observed on a number of occasions in Kenya, Zambia, and Zimbabwe. Aggression was mostly by a grazing adult against an individual egret in motion on the ground, at a distance of 0.5 to 1.5 meters ahead of the buffalo. The buffalo charged the bird without any previous threat display but often initiating the behavior with a grunt, head low and hooking with a horn, all with much violence. The egret always successfully avoided the attack, usually by becoming airborne for a moment, then landing a short distance away. On two occasions the egret failed to leave the ground at once and ran or skipped out of the charging buffalo's way. The buffalo then continued the pursuit for a few seconds, searching with the horns. Aside from this, however, all indication of aggressiveness disappeared immediately after the attack.

I interpret this behavior as a response to an unidentified moving nearby object – a potential danger. Previous to the attack the grazing buffalo's vision is partly hindered by grass between it and the egret, complicating identification. Furthermore, it seems probable that buffaloes are generally incapable of positively identifying very near objects by sight alone. The behavior is a noncognitive response to possible imminent danger.

Minor predation

Hyena

Syncerus c. caffer adults are too powerful when in optimum condition and their nonadults usually too well protected to make a worthwhile prey for hyenas (*Crocuta crocuta*), leopards (*Panthera pardus*), or wild dogs (*Lycaon pictus*). However, when prey is scarce or there is a special opportunity, predation of the buffalo is at least intended by these species and, at least in the case of the first two, may be successful occasionally.

In the Lorgorien area, Kenya, in 1965, I observed a solitary male buffalo being followed at some 10 meters by four spotted hyenas. The hyenas appeared intensely interested in the buffalo. The buffalo turned to face the hyenas and stood motionless watching them for 2 minutes while they stood or moved sideways watching him. Then the buffalo walked away, the hyenas following at a greater distance. I saw the same male buffalo alive several weeks later. At another time, on the southern slope of Lolgorien's Kimoigoyen Hill, in 1965, a large male buffalo was shot while charging and finished in a kneeling position, with the head unusually high and level. It was in the evening and the dead animal was left undisturbed overnight. Hyenas were seen approaching the spot at dusk, and their vocalizations were heard at about 2200 hours. Well after sunrise on the following morning, I observed four hyenas near the dead buffalo. There was only one minor laceration on the buffalo, which appeared very recent. If the hyenas spent the night near the dead buffalo, an explanation is necessary why they did not feed on the carcass for some 10 or 11 hours. One explanation is that they were exercising caution toward a large buffalo in a plausibily lifelike posture, though motionless. These two observations are compatible with a hypothesis that the Lolgorien hyenas looked on buffaloes as prey, but feared their response to predatory aggression.

On March 13, 1980 C. S. Churcher (personal communication) observed a cow buffalo killed by hyenas. Churcher was at Treetops Game Lodge as dusk was falling and observed a group of about 20 buffaloes approach the waterhole from the north. The group may have comprised a basic herd as there were only 3 or 4 adult males, a number of adult females, many with calves, and immatures of both sexes. The group watered at the north side of the pool. While the buffaloes were watering 2 hyenas approached from under the lodge and moved toward the group, skirting around it near the forest margin. The hyenas behaved as though they were testing the buffaloes but the buffaloes stayed in a compact mass with the older cows and bulls on the edges and resisted the hyenas' testing dashes toward them by snorting, head tossing, and advances with heads lowered of a few steps before backing into the compact group. After 5 or 10 minutes of this behavior the pair of hyenas was joined by another and the 3 hyenas then started to press the buffaloes to move toward the lodge. The buffaloes were moving slowly in this direction while the hyenas were increasing the frequency of their dashes and becoming bolder. At this point it became apparent that they were concentrating their attention on one mature cow without calf. She seemed to be in good condition, uninjured, and not senile, but did shake her head in a twisting motion as though there was some ear irritation.

Quite quickly and all at once the buffalo group moved toward the lodge and assembled beneath it, but the cow to which the hyenas had been paying attention remained behind. She began to chase the hyenas in front of her, attempting to use her horns, while the hyenas behind her and to the flanks darted in to nip at her tail, legs, and belly. Shortly after she was isolated, a fourth hyena joined the others. The hyenas were making little noise although they had called early in their worrying of the group and had been answered by others in the forest.

The 4 hyenas began to worry the cow in earnest, snapping at her from all sides, and even resisted her short charges and snapped at her face. Her charges were not as long as earlier and she had been nipped in the flanks and tail, and she appeared to be becoming tired. By now the light had faded so that the drama was that of dark shadows in a lethal ballet between the edge of the forest and the pool, and most details of movement were unobserved.

Additional hyenas arrived during this stage, but the light was too poor to determine how many. They bit at and hung on to the cow, and coerced her toward the forest. At no time did she try to rejoin her group, which waited near the lodge for most of the drama, before slowly moving away. No member of the group attempted to help her, nor was there any behavior displayed beyond interested staring, head tossing, and the older bulls and cows remaining between the calves and the hyenas.

Churcher did not observe the killing of the cow, which occurred at the forest margin, nor was the cow observed to go down. However, after she had been killed and the hyenas had commenced to feed on her, the eyes of those that faced toward the lodge were visible as they reflected in the lights that illuminate the saltlick, and it appeared that about 6 individuals were feeding on one side. Thus about 12 or 15 hyenas may have been involved at the end.

In the morning, the game scout took a party, including Churcher's son and two daughters, to see the site of the kill. They reported that only the skull, axial skeleton, pelvis, and one hind leg remained. The other elements of the skeleton had either been eaten or carried away, and all the soft parts except the brain had been consumed.

It is quite likely that predation by hyenas has increased in recent times, as compared to scavenging. Hyenas commonly scavenge lion kill remains. However, the number of lions, and therefore lion kills, has become much reduced in the last few decades. The predatory behavior of hyenas may have been increased in response to this scarcity of lion kill remains to feed on.

These observations are cited because they add weight to those of Kruuk (1972) and Sinclair (1977), which also suggest that buffaloes may occasionally fall prey to hyenas. However, Kruuk and Sinclair both mention much greater numbers of hyenas (11 and 69) in cases of intended or apparent predation on buffaloes.

Leopard

In July 1974 in Zambia, I was resting overnight on the western edge of the Kasolo Kampinga thicket in the Busanga flats. At 0115 hours I went to examine a buffalo herd that I had been observing until after dusk. I heard the buffaloes and approached quietly over sandy ground. A moment after sighting the buffaloes some 100 meters away from the thicket, I saw movement unconnected with the buffaloes, and with binoculars identified a large leopard. The leopard advanced on the buffalo herd. Twelve minutes later (0204 hours) there was commotion among the buffaloes which stampeded. I became aware that a buffalo was down, and soon deduced that it was being pulled along the ground. At dawn, the carcass of a female calf some 4 to 5 months old was found, wedged in the angle between the steep slope to a termite mound and a small tree that grew on it. The kill had been dragged about 75 meters, with two halts along the way. The neck region was lacerated and some of the vertebrae dislocated. A small part of the front quarters had been eaten but the kill was otherwise quite intact.

Crocodile

I do not have direct records of crocodile predation on buffaloes. Possible predation by crocodiles (*Crocodylus niloticus*) is limited by the smallness of the overlap between the two species' habitats. Buffaloes were observed to avoid water where crocodiles abounded, including some river or lakeshore sectors and large pools in marshy country. However, in the same localities access to water was often less easy than elsewhere, and hippopotamus was often present. Thus the reason for avoidance is unclear, since all or any of these factors could contribute to it. Buffalo groups fording parts of the Lufupa drainage, western Zambia, where large crocodiles were common, were often observed to move fast, in unusually compact formations. This behavior was an efficient deterrent to crocodiles, regardless of its actual cause. One adult female buffalo carcass in the Uaso Nyiro River, Kenya, 1967 and one juvenile carcass in the Busanga section of the Lufupa drainage, Zambia, 1977 appeared to me, on circumstantial grounds, to have been crocodile kills. Pienaar (1969b) mentions buffalo as falling prey to crocodiles in the Kruger National Park, South Africa. The Uganda Game Department reported in 1947 that a crocodile succeeded in overcoming a buffalo in the Busoga area (in Cullen, 1969). That very large animals can succumb to crocodiles of the nilotic species, and that therefore some predation on buffaloes can be expected, is apparent from a letter that M. C. Fleischmann wrote in September 1907 to President Theodore Roosevelt, describing the taking of a fully grown female black rhinoceros (*Diceros bincornis*) by one or more crocodiles in water, near the confluence of the Tana and Thika rivers, Kenya. The description is substantiated by three photographs which, together with the letter, were published in Selous (1908).

Aggressive behavior directed at vegetation

It was observed in Kenya, Zambia, and Zimbabwe on more that seven occasions, in male buffaloes rubbing their horns on small trees, that this behavior sometimes acquired all the appearances of open aggression. The buffalo emitted grunts, his head movements became violent, and he tried to push the tree forward and down, all the way to the ground if possible. Once in Zambia, in 1977, I observed a male which was rubbing his flank against a sapling, turn on it violently, and "fight" it. Buffaloes were never observed to "fight" thick trees against which they rubbed. It appears possible that the resilience of a small somewhat pliable tree against the buffalo, as the latter repeatedly exerts pressure on it, simulates live resistance and serves as a release mechanism for aggressive behavior. Pressing from the front against the forehead or boss with a hand, tried in the Antwerpen Zoo, Belgium, 1976, and Munda Wanga, Chilanga, Zambia, 1978, resulted in the buffalo, in both cases of the *caffer* subspecies, instantly to push back with some violence, as is common in many bovids.

II. Intraspecific encounters

Intensity levels

Intraspecific agonistic encounters of *S. c. caffer* vary in intensity according to their precise function. High-intensity encounters, which involve reciprocal escalating conflict, are the most costly and by far the least common. They tend to occur between two individuals so close in status that criteria are lacking by which one will concede precedence to the other, and only violent confrontation remains. The winner retains the disputed position, while the loser drops in status. Such high-intensity encounters were observed predominently between males. The restricted number of functional positions for males within buffalo herds leads to increased competition and favors male hierarchies that are mainly linear. The linearity of a hierarchy itself becomes a reason for high-intensity competition, which thus may occur in isolated bachelor clubs, without any females immediately available. However, within bachelor clubs, the linear rigidity of the male hierarchy may occasionally slacken, as is apparent from the observed temporary triad relationships within male groups (Chapter 7). Females may use horns intraspecifically, but not nearly as often as males. The three instances that I saw had the character of a rapidly escalated and deescalated flare-up.

Where some ranking criteria exist between two individuals but the net balance in status is small and occasionally appears uncertain, medium-intensity rank-enforcing encounters occur, which may end in one of two ways: either the higher-status individual impresses his or her superior position on the interloper, who then deescalates the conflict, or the interloper stands fast while the other retires. The latter alternative may occur when the originally higher-status individual is reduced to a lower rank through sickness, injury, advancing age, sexual depletion, or loss of shared status (Chapter 7; discussion of "borrowed" status). Rank-enforcing encounters may occur between individuals far apart in a hierarchy, as when a mature male chases a much younger one. This behavior was observed to result in a transfer to one of the bachelor clubs of those subadult males that were acquiring adult character but had not yet left their basic herds. Within a bachelor club, an individual enforces his position, by threat or open aggression, relative to any other club member that may interlope. This is usually deescalated, mostly by the interloper, who may flee immediately or concede after some resistance; or by the original aggressor, if he feels bettered. Rank-enforcing encounters are unlikely to cause the loser to drop several rungs in the order, although a large drop seems to occur once in a while, through repeated rank enforcement against one individual. Enforcing encounters occur between calves and females, usually the mothers, when they do not wish the calves to suckle them, occasionally or permanently. These females may turn their bodies to place the udder out of the calf's reach, or hit the calf on the flank with the head or a horn. Occasionally these blows may be so hard as to make a hollow sound on impact. Limited combat, or head sparring, is a common type of medium-intensity encounter, in which individuals test their strength and ranking relative to that of close associates.

Low-intensity agonistic encounters have a confirmatory character. They are displays of superiority or submission, readily accepted by both participants of these interactions. Rank-enforcing and confirmatory encounters were observed between members of both sexes, between monosexual and heterosexual pairs. Sexual behavior leading to copulation is discussed in Chapter 9, but may be considered a form of agonistic encounter.

High-intensity combat

I have observed nine high-intensity combats between pairs of adult males, in their entirety or almost so. Four were in Kenya (Lolgorien, 1965: Aberdares, 1966; Mt. Kenya, 1974), three in Zambia (Busanga, 1977; Chunga, 1978), and two in Zimbabwe (Mabalauta, 1972).

When two bulls recognize each other as competitors, they remain some distance from each other and display reluctance to give ground. Then they circle, possibly to outflank one another, which they are usually unable to achieve, or to make a prolonged lateral display toward the opponent. I believe circling to be primarily a display of size and implied mass, intended to give either belligerent the time to recognize his own inferiority and retreat. Circling is accompanied by other threat displays that include head tossing and its higher-intensity equivalent, horning and tossing of shrubbery and soil. The observed duration of these preliminaries varied from 7 to 11 minutes. If neither individual retreats, both ready themselves for an instant reciprocal response to a head-on charge. If there is plenty of open space, they may charge each other across as much as 30 meters according to Sinclair (1977), although the distances observed by me were somewhat less. Buffaloes do not charge one another at their top speed, partly because the distances involved and some degree of ground unevenness may prevent the necessary acceleration, and partly because the momentum at impact, if the combatants were large bulls of some 800 kilograms, would be approximately that of a medium-sized four-passenger car hitting a house at 48 to 56 kilometers per hour. My own best estimate of buffaloes' charging speeds is 13 to 22 kilometers per hour. If the bulls charge each other at 20 kilometers per hour, the momentum at impact is equivalent to the above car hitting a house at 23 kilometers per hour. As the animals collide, the blow on the head is cushioned by a rotational movement whereby the chin is tucked farther back, and the dorsal part of the neck is stretched downward and forward, while the bulky thoracic portion of the body bears forward over the head. The resultant great strain on the upper neck and dorsothoracic assemblage of muscles may explain their exceptional development in the heavier races of *Syncerus*. At the climax of a head-on charge, therefore, the heads are very low and the withers high and riding forward. The recovery from this position depends on the ability to push forward and upward with the head, against the opponent. This slight but strenuous movement reestablishes balance and gives a final advantage, because the other, weaker individual is blocked from moving forward, although he must keep in motion to retain balance; so he moves sideways, thereby exposes his flank, and runs to avoid injury. This escalates into outright flight, with the winner in pursuit for a few tens of meters. If initially the combatants circle close to each other and charge across a few meters only, the combat results in a fast and loud pushing and horning match, largely composed of improvised moves. It may continue for several minutes and is the most likely kind of encounter to end in serious injury. The head-on charge that lasts only a moment may also, however, result in substantial injury. I watched a high-status male have his horn snapped off at the place where it thickens to form the boss, during a head-on charge, in the Busanga flats, Zambia on

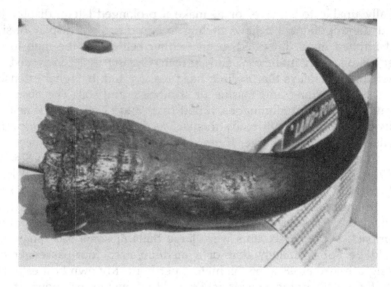

Figure 8.6. A mature male's left horn broken off by impact during a violent encounter between two high-status males. After the fight, the horn's owner had difficulty in maintaining balance or walking in a straight line for several days. The horn remained attached by a strand and fell off after 11 days. (Busanga area, Zambia.)

August 12, 1977 (Figure 8.6). The horn remained attached by a strand of tissue, dangling next to the bull's left ear. The animal was pursued for only some 20 meters. Shortly afterward I saw that he was reeling and then lay down. The bull remained on his own until August 21, when he rejoined the herd. Throughout this time, he clearly lacked balance, reeled, staggered, and lay down frequently, whenever seen. On August 23, the broken horn dropped off, in my presence. An inspection did not reveal any previous damage. The injured male appeared to regain his health, continued to associate with the high-status males, and retained contact with them until October, when he strayed away.

Limited combat or head sparring

The main purpose of limited combat in *S. c. caffer* appears to be a forceful, but mostly injury-free, statement of an individual's place within a ranking system. Limited combat is most common where ranking is most subject to modification, that is, among immature individuals no longer attached to the mother. As they grow, their strength changes rapidly, both in the absolute and relative sense, and necessitates frequent testing. Since a buffalo herd's demand of males is usually less than the supply, ranking competition among them, including combat, is more pronounced than among females. The need for contests decreases as buffaloes attain maturity, when their physical attributes change less rapidly and ranking within the hierarchy is more permanent. Contests are not eliminated, however, and limited combat occasionally takes place between mature adults.

The combats are basically head-pushing or head-sparring encounters. The younger the participants the greater the likelihood of dissimilarities between them, in size and other characters. Combative behavior occurs even in calves, which may occasionally challenge an adult, often their mother. It is important to note that young calves, faced with *extraspecific* threat, have been observed to charge violently. However, intraspecific combativeness in calves, juveniles, and young subadults has the general appearance of play. The following field note describes an adult–young subadult encounter, on the morning of July 20, 1974, in the northeast corner of the Kafue National Park, as I watched a herd walk slowly in an extended line along the Kebumba *dambo:*

> A young male subadult is walking some 5 meters ahead of a fully mature bull. Suddenly the subadult begins to prance and lower his head. He turns around and faces the bull. The older animal lowers his head without haste, halts, and the two fit heads together. The mature male starts to walk forward, easily pushing the straining subadult in front of him. Then the subadult prances away, to restart the contest a moment later. This is repeated four times. Afterwards, both animals resume ordinary walking (07/20/74/0910).

Head sparring is most often observed between individuals 2 to 4 years old. Encounters within this age class, and also between mature males, begin formally. The combatants tend to be similar in size, and frequently have been standing or lying near each other immediately before the contest. The opponents take up positions facing each other at close quarters, with deliberation, then try to place the boss of the horns in contact with and below the opponent's boss, and shove forward with the body and upward with the head. When one of the combatants is unable to keep his head down, he backs off. This either constitutes the end of the contest or more rounds follow. The maximum observed number of rounds was eight, lasting a total of 4 minutes.

The frequency of limited combat decreases with age, after reaching a high in the later subadult stages, but the intensity tends to increase, and contests between mature males may be so violent as to result in injury, often to the horns. This escalation does not appear to be intended, and the loser, at worst, exchanges his status with the winner, but his life is not otherwise affected. It can be determined by the preliminaries whether a combat is a limited or a high-intensity encounter, since the latter begins with circling, whereas the former normally begins without it. Hooking with a horn, but with heads kept fairly close together, occurs without great force in subadult encounters, but more violently in some adult ones. Limited combat may be accompanied by a few vocalizations, mainly clipped ones that sound like "ba." In most cases, limited combat is distinct from any other behavior but in a few cases it grades into high-intensity combat or into low-intensity agonistic behavior. The following three field notes illustrate cases where limited combat encounters did not work out in the normal way. In the first case, there was a mismatch:

> A pair of subadults have a 2-minute match. It starts in the normal way, then one begins to push the other with relative ease. Finally the losing animal tries to push with his flank – any old how – then gives up and walks away. The winner resumes grazing (04/25/73/1654).

In the following two cases, contests failed to materialize:

> The subadult males go through the routine of searching for each other's boss, then one places his head on the other's back and both walk on together (10/17/72/1720).

> Two adults face each other. They assume threat postures for a second, then relax. One of them falls behind the other and they walk away (01/08/78/1605).

The following two notes illustrate limited combat between mature adult males:

> Two large bulls fit their bosses against each other and start pushing; after about 30 seconds one backs up a little; the winner tries to reengage but the loser keeps avoiding him by moving his head out of the way. Eventually, the loser walks slowly away (10/23/71/1728).

> Two large bulls from the center group get up with a single bellow, stand away from the group, and have a single round of violent combat, the loser retreating at a run back into the group. This causes some disturbance: three bulls stand up looking outward (away from others) in aggressive postures (10/26/71/1618).

Limited combats between subadult and young adult females were observed, in the Antwerpen Zoo, Belgium, in 1976 as well as in wild African habitats. They are much less frequent than between males. They tend to be shorter, and have never been observed to approach escalation toward violence, as was the case with some male contests. In a sample of 105 daylight observations of limited combat, 24 percent occurred between 0700 and 1200 hours, while 76 percent occurred between 1400 and 1900 hours. Such combats occur also at night but are fewer than during the day and their distribution could not be reliably determined.

Sinclair (1977) uses the term *sparring* when commenting on some of this behavior as observed in the Serengeti National Park, Tanzania. *Head-sparring* is used by Hafez and Bouissou (Hafez, 1975) when they speak of head-to-head pushing contests in domestic cattle.

Deescalation of combat

The "clinch" is a well-known pattern of agonistic interaction in domestic cattle. Hafez and Bouissou (Hafez, 1975, p. 228) describe it as follows:

An interesting pattern of behavior, the clinch, occurs in prolonged fights between females; one participant allows the opponent to gain a flank advantage as it pushes its muzzle between the opponent's hind leg and udder (Schein & Fohrman, 1955). Neither cow can attack from such a position and should one attempt to turn for an attack, the other rides along by maintaining body contact. The fight cannot be resumed until both participants are ready. The clinch is an ideal way for combatants to rest safely during prolonged bouts and is not seen often, since the majority of fights are decided quickly.

In African buffaloes of the *caffer* subspecies, a similar pattern occurs in adults of both sexes. Where observed, however, it is not a means of resting during prolonged limited combat, as in cattle, but of avoiding combat. As one individual approaches another from the side, either submissively or threateningly, the recipient individual maneuvers itself into the clinch posture. By then the first individual has assumed an analogous posture, and the two of them turn around at a rapid pace, maintaining flank-to-flank contact and chasing each other's rump. The behavior may become modified by resting chins on rumps. This could be an accommodation to the encumbering horn shape. The behavior has been observed to continue for up to 50 seconds, and combat has never followed. The interacting animals were always very similar phisically. I interpret this behavior as a conciliatory response to a possibly misunderstood signal.

An interesting case of an imminent combat that was deescalated through intervention occurred in the Busanga area, Zambia, in the large bachelor club described in Chapter 7. I transcribe the field note of the event:

Bull A is ruminating quietly; bull B comes to him with head lowered; A lowers his head to engage B but B sidesteps so that the two of them are going in a circle, one following the other. They are slowly making ready to meet heads when the most dominant bull comes over, sends B on his way, then muzzles A gently, which is reciprocated with a couple of gentle muzzlings, then the dominant bull walks away (10/27/71/1717).

The initial circling indicated that the challenge was serious. Bull A was evidently ready to make little of the encounter but B wanted to escalate. The two were high-status bulls, very close in ranking, with A slightly ahead of B. To judge from the observation of other similar encounters, this one promised to become a high-intensity combat. The dominant bull's interference had every indication of being intentional. Previous to this interaction I had not noticed during 5½ days of observation any special relationship between the dominant bull and A.

One-sided rank enforcement

If a subordinate gets in the path of a dominant, the latter responds aggressively. The aggressive response involves horn display by tucking in the chin, which may be accompanied by lowering the head and hooking, head tossing, and a silent or grunting advance toward the subordinate. If the margin of dominance is wide, the subordinate retreats at the first sign of threat display. However, the situation arises most often in moving herds, when a subordinate blocks the way of a dominant unintentionally, by walking close in front of the latter. The dominant individual may then grunt and take a run at the offender. If the offender fails to get out of the way quickly, he or she gets butted in the rump. If a subordinate animal blocks a dominant's path more than once within a short time, the dominant intensifies the chase and makes a point of butting the offender, usually in the hindquarters.

As the margin of dominance narrows, the subordinate's response varies accordingly. The threat may be returned, although less strongly displayed, sometimes only by a reluctance to give ground. The animals may even start to circle each other, but the subordinate is likely to retreat in the first few seconds. Alternatively, the dominant aggressor charges with head low and a grunt, causing the other to flee. If the recipient of the charge does not have time to flee but is compelled to make contact, he or she disengages as fast as possible and flees. The loser may be chased for a few tens of meters. Sometimes the recipient reacts very slightly to a threat, perhaps merely turns the flank to the other in a lateral display, and moves without haste. Slow movement and reluctance to depart from another's presence are important agonistic behaviors. The only sign of stress may be the absence of grazing, ruminating, or resting behavior when these would ordinarily be expected. This may happen when the difference in status is small, and then there is a chance of escalation. Quite often, however, the recipient walks away, preventing escalation. It must be emphasized that in the majority of instances one animal rapidly recognizes another as his dominant and consequently no dispute arises. Uncertainties as to relative ranking occur between individuals that are not close associates, do not know each other intimately, and fail to identify any characteristic that would immediately rank one of them above the other. If the en-

counter occurs amid animals closely associated with one of the interacting individuals, while the other is a relative intruder, though a member
of the same herd, something akin to territorial behavior occurs and the
intruder does not press for a confrontation: Such interactions do not go
beyond lateral or head-on displays and a reluctance to give ground. All
the high-status males, however, and many high-status females are immediately recognized throughout a herd, and the largest individuals
are readily allowed precedence.

Low-intensity interactions

Low-intensity encounters in which mild coercion may be used to confirm an individual's dominance over another, but never violence, are
frequent. One common means of advertising dominance is to compel a
recumbent subordinate to stand up, as in the following instances (field
notes):

> A mature male comes up to the subadult male that is lying down,
> muzzles him gently for about 30 seconds, pushing him equally gently
> now and then with head and neck; the subadult gets up and stands by,
> while the other lies down on the subadult's previous spot
> (10/22/71/1230).

> A male is lying down; another approches him from behind, muzzles
> him, and pushes him gently with the boss, until he gets up, then the
> aggressor relaxes (01/08/78/1603).

Sometimes the recumbent individual will not be persuaded to stand
up, but this need not generate disharmony; the dominant individual
may settle for less:

> Bull A comes to bull B which is lying down, and gently pushes him
> with the boss of the horns; B does not move; A rests his head on top of
> B (10/22/71/0930).

When two individuals interact on this level, the more dominant one
tends to place its head over the head, or some other part, of the
subordinate, as in the preceding and the following field note:

> The number-1 male reverses the direction and starts to graze up-
> stream. As he passes by the oldest bull he leans against the old bull's
> side and rubs, then puts his head over the old bull's, which then gently
> moves his head up and down. Then the number one bull walks on,
> followed by the old one (10/25/71/1755).

Conversely, an animal advertising subordination places its head beneath the dominant's chin or some other part of its body:

> One of two bulls starts rubbing the other under the chin with the
> head and shoulder, the other resisting gently. The rubbing becomes

flank to flank, so strong that it makes the bulls "bounce off" each other, touching only with their hindquarters (07/16/74/1737).

Subordination may be advertised by a behavior described fully by Sinclair (1977, p. 111):

Submission is indicated by a head position contrasting to that in threat, with head held low and parallel to the ground, horns lying flat and back sloping down to the hindquarters. . . . A submissive male in this position will actively approach a threatening male and place his nose under the belly of the dominant, . . . sometimes under his neck or between his back legs. Immediately afterward the submissive male will turn and run, and at the same time emit a loud bellow. When bellowing the mouth is noticeably wide open . . . with tongue protruding and curling up. Submissive males sometimes have their mouths open even before they have reached the dominant and are beginning to make their bellow, although at this stage it is no more than a low croak.

Occasionally, the broad observation that the more dominant animal's head is placed on top, or the more subordinate animal's underneath, depending on which of the animals is active and which passive during an agonistic interaction, appears subject to exceptions, when individuals that one would expect to be subordinate place their heads uppermost:

A 3-year-old approaches a mature bull and starts muzzling him; older animal reciprocates. The subadult rubs his chin on the adult's boss and they rub cheeks, carefully fitting the adjacent horns to make cheek contact possible, and rub sides (04/25/73/1532).

Wallowing and its role in agonistic displays

It is thought worthwhile to stress here, with an apology to readers who may find it excessively elementary, that *Syncerus* is not a "water buffalo," and neither wallowing nor entering water is a habitual behavior, as it appears to be in *Bubalus bubalis*. The African buffalo tends to favor marshy areas (though it dislikes resting on wet ground) and it is possible that, had there not been so many crocodiles in most larger bodies of water in Africa until quite recently, *Syncerus* would have retained or developed water-entering behavior on a daily basis. It is a good occasional swimmer, able to swim out to islands in lakes (e.g., Pitman, 1942) and across rivers. However, the genus inhabits and grazes regions that, at least seasonally, may have very little surface water. We may speculate that seasonal droughts, crocodiles, and local highland habitats have combined to modify *Syncerus*'s watering habits. In any case, whereas all African buffaloes queue up to drink on a daily basis, with only rare exceptions, they do not treat wallowing in the same way. Even when there is ample opportunity, only some, or none, may wallow. Individuals of both sexes and various status wallow occasionally. I have observed several animals of both sexes wallowing together, in fairly restrictive mudholes, in Zambia and Zimbabwe. I have seen a few times whole sections of large herds wallow simultaneously upon encountering in their path a suitable pool near the edge of some Zambian *dambo* in the rainy season, but this was incidental rather than habitual behavior. In those months when facilities for wallowing are reasonably good, its cooling effect is relatively unimportant, since the weather is not especially hot. It is during the dry season, when there is no mud, that the need for cooling off may appear. Wallowing serves to reduce the biting of flies and ticks. I noticed ticks imbedded in pieces of mud that had dried on a buffalo's body and then spalled off, although I doubt the large-scale effectiveness of this process. Thus wallowing appears to be mildly beneficial, and probably pleasurable, but not a requirement. Sinclair (1977) attaches much importance to wallowing as a social function, that is, a complex threat display, of high-status males.

In general, high-status males tend to stay by themselves more than the other herd members, whether they rest, graze, drink, or walk. It is therefore not surprising that they are relatively often seen in wallows by themselves. Occasionally, though not always, solitary wallowing by high-status males is accompanied by agonistic-appearing behavior. To illustrate, I again transcribe a field note:

> All buffaloes have passed through the small water/mud hole. Only Trailer (number-1 male) stops and takes a mudbath. He wallows on both sides, legs up. Then getting up on hind legs but kneeling on front legs he scoops mud, grass etc. with his head and tosses it, then again; then repeats it all several times, for over 3 minutes, then he *runs* after the herd (11/18/77/1228).

This is a curious behavior. It is often done where it cannot be noticed by other herd members, or other species, as far as the buffalo can know. I have never seen or heard of it being done ritualistically in dry mudholes, and I have observed that its performance is erratic in wet ones. Perhaps it is not basically an agonistic display. Large males of the *caffer* subspecies have a skin fold underneath the back edge of the horny boss. The neck skin which lines the fold is relatively thin, and it is a place where ticks tend to collect. The "horning display" may be simply a way of scooping mud into that fold, as a tick and tsetse fly reduction exercise. The angles and movements observed by me are compatible with such a purpose. It is possible that the movements, which in general are associated mainly with combat, act as a releaser for additional agonistic combatlike motions. However, I am disinclined to consider wallowing in *S. c. caffer* as an important agonistic display.

Concluding comment

Intraspecific agonistic behavior of S. c. caffer grades from openly violent and potentially deadly, which is rare, by way of moderate, where violence is a recognized last resort and tends to be readily displayed, to mild, where interacting animals demonstrate their acceptance of their ranking by gestures reminiscent of conflict but devoid of violence. The higher-status individuals are capable of behaving at all these levels, which are not mutually exclusive, and gentle displays toward subordinates were observed. S. c. caffer is influenced by body size during intraspecific exchanges, probably because mass, more than agility or speed, decides its high-intensity agonistic encounters, and the lateral display, which best advertises an individual's bulk, is therefore favored. Head butting, which the African buffalo shares with other bovid genera such as *Ovis* and *Ovibos*, is a noteworthy behavioral development in a form as massive as S. c. caffer.

III. Various agonistic behaviors

The alert stance

This stance, one of the most characteristic of S. c. caffer (Figure 8.3), is adopted whenever a buffalo's attention is attracted by any event that may prove to contain danger. It serves thus as a signal of possible danger. It is also adopted ritually, when confronting conspecifics of no close affiliation, notably when two herds meet. References to the alert stance are made elsewhere, when describing particular circumstances in which it occurs.

When a group of buffaloes is alerted, all individuals do not normally face in the same direction when they assume the alert stance, but orientate themselves so that most of the surroundings are under surveillance. This is the common pattern even in groups of three or four.

The appearance of other nonpredatory species may elicit interest in some group members, which assume the alert stance facing the newcomers. The following field notes from Zambia illustrate:

> As a group of wildebeests walk abreast of the resting buffaloes some 150 meters away, two bulls and three cows of the nearest cluster stand up and scrutinize at the alert, then start to lie down again after 50 seconds (all are down in the next 110 seconds). Others remain inattentive throughout (08/25/77/1113).
>
> Pathfinders on the right wing at the alert toward eight wildebeests walking in file about 500 meters away. They retain the alert stance for about 90 seconds, then resume grazing more or less together (10/30/77/0706).

The passage of a strong gust of wind often elicits a hasty assumption of the alert stance in separate individuals or several group members, who then do not necessarily assume the stance facing into the wind but may face in several directions. The sudden change in the airborne odors reaching the buffaloes at that moment may in itself be the stimulant, even if no particular threatening scent, for example, lion or man, is actually discerned by them. The behavior may contain an allelomimetic component.

An unexplained agonistic behavior

I transcribe a field note (10/23/71/1735) taken in the Busanga flats, Zambia while observing a large isolated bachelor club:

> While the majority of the bulls is down or otherwise at rest, two are 15 meters to the side of the group. They appear to start walking toward me, then stop, looking as if at me for about 1 minute; then they start following one another in a small circle (diameter about 7 meters), counterclockwise, apparently sniffing near the ground; every time one is in position X, he "jumps" as if to run but does not do so. After a few circlings they mingle with the group and all remains peaceful.

Agonistic behavior in captive individuals

Five of ten adult African buffaloes observed for 27 daylight hours at the Antwerpen Zoo, Belgium in September 1976 were of the taxon *S. c. caffer* var. *capensis/radcliffei*. Four of these, two males and two females, the oldest (a male) about 9 years of age, were repeatedly observed to bite the horizontal metal bars of their outdoor enclosures, usually for a few seconds at a time. It was very apparent that the behavior was habitual. One male (about 5 years old) bit the vertical bars systematically, one by one, the length of the outer enclosure. The type of enclosure can be seen in Figure 2.6. Biting behavior is known among domestic ungulates under stress from confinement, notably horses (Fraser, 1974). The fifth adult individual, not observed to bite enclosure bars, was a female with a suckling calf. These animals were at the time kept together in pairs of the same sex or separately, in a row of adjacent enclosures. The biting behavior occurred in either case.

The two bar-biting childless females, when together, were observed on one occasion to engage in a series of four typical head-sparring encounters. The younger adult male, mentioned above in connection with vertical bar biting, tried to engage a bison (*B. bison*) in an adjacent enclosure in a head-to-head combat. The bison cooperated and the pair engaged as closely as was possible through the bars, then disengaged and moved apart. The same male twice displayed a typical alert stance toward a human passerby.

9

Reproductive, mother–calf, and nonadult behaviors

I. Reproductive behavior

Precopulatory behavior

The reproductive sequence in *S. c. caffer* begins with a male preparatory phase whose function is the selection of males suitable for reproduction. This is achieved through male agonistic encounters (Chapter 8), which serve to erode the confidence of the losers and build up that of the winners. A state of orderliness results, so that when a female enters estrus, she is neither rushed by all the males at once, nor is there need to have elimination contests between them as she waits: Instead, the lower-down males automatically give way to higher ones. An adult male is not restricted from mingling with females by means of some quantum difference but by the appearance on the scene of a higher-status male. The lower-status male takes a "waiting" position down the line. Thus the strongest males have the first choice. The system operates on a continuous basis and the higher-status males can time their entrances as they choose. When they go away, lower males may arrive shortly thereafter; when they return, the lower ones retire. If they fail to return, the lower ones are still there to fertilize the females. The lower-status male, while preoccupied with an estrous female, may not immediately realize the arrival of a higher male. The latter may have to remind him by a rank-enforcing grunt and butt. The system works well, however, and the other hardly ever resists. Thus the male linear hierarchy conditions all buffalo bulls into being efficient potential breeders.

The second phase begins when a female is recognized as approaching estrus. The approach of estrus causes tension in the *Syncerus* female, as it does in *Bos*. In *Syncerus* it may be manifested by an occasional abruptness of movement, sudden assumptions of the alert stance, and little

180

sideways head jerks originating at the top of the neck. As in *Bos*, recognition of the beginning of estrus by a human observer is largely a matter of guesswork. In retrospect, I guessed that one Busanga female, 1974, may have been in estrus for 63 hours or longer, and a second one, at Chunga, 1978, may have approached that length of time, but in other cases, the duration appeared to be 1 or 2 days, which also appeared to be the case at the Antwerpen Zoo, Belgium, 1976.

Many hours before the full onset of estrus, herd males begin to detect it, mainly by olfactory means, while routinely sniffing females in the region of the vulva, or from grass on which the female may have lain or urinated. In the latter case, the male may suddenly start inspecting females, walking energetically from one to another. On locating an estrous female, a male stays by her side, that is, "tends" her as described for domestic cattle (Hafez, 1975). Tending is not constant in the early stages of the onset of estrus. After the first discovery of a female in early estrus, and initial tending, males may lose interest. Other males that visually recognize the tending behavior may come up and replace the first animal.

Flehmen is an early response of the male to the estrus scent and is likely to accompany his first discovery of the estrous female. This widespread and prominent ungulate behavior consists superficially of a forward extended neck and muzzle, the upper lip curled up, exposing the gums and teeth. The nares are described by Hafez and Bouissou (Hafez, 1975) as "distended" in domestic cattle, and they are similarly described by Sinclair (1977) for *Syncerus*. My several observations of flehmen in the *caffer* subspecies of the African buffalo indicate constricted nares. Dagg and Taub (1970), describing the behavior in species other than *S. c. caffer*, state that the muscles used in producing the flehmen conformation together serve to restrict or close the nasal apertures, while Fraser (1974) draws attention to the nostril constriction in stallions (*Equus caballus*) during flehmen. There is a wide agreement that flehmen is a response to olfactory stimuli, although its exact function is queried. In *Syncerus*, the frequency of flehmen does not appear unusual, as compared to other ungulate genera in which it is known, but when it does occur it is often very pronounced. Whereas in *Bubalus bubalis* the line of back, neck, and head during flehmen may be more or less horizontal, in *Syncerus* the head is often elevated. I observed flehmen in five single or paired bachelor males, in advanced maturity, almost certainly nowhere near a recent path of a female, and certainly several kilometers from any herd (Lolgorien, Kenya, 1964-65; Ruaha National Park, Tanzania, 1973). I feel, therefore, that it is possible flehmen is occasionally used by *S. c. caffer* during general olfactory probing of the surroundings, unconnected with sexual behavior.

When a female becomes well established as one approaching estrus, the tending by males becomes more earnest and the replacement of one tending male by another definitely acquires the character of dislodgment. Chin-resting on the female's rump becomes more frequent, and again resembles the analogous behavior in domestic cattle (Figure 9.1a). Until she is ready to receive the male, the female either moves away from under his chin, which rests on her rump, or stands, but without leaning into the male. The male may mount her at this stage, but without completion.

Copulatory behavior

The next phase is that of full estrus. The female is tumescent and very possibly attracts males within a wider circle than previously, and the increased intensity of her scent may be a signal to the males. The highest-ranking unengaged male within her range responds by dislodging the animal tending her to date. When a male chin-rests a female at this time, she stands wide and leans into him. The male mounts her in the manner of a domestic bull, that is, the withers above the female's pelvis, front limbs straddling her, and his hind limbs barely touching the ground behind her as his abdominal muscles contract. During the mounting, the impression is that of considerable motion, the male sometimes shoving the female forward while she braces herself under his weight. The performance is often watched by other males, in alert or threat postures (Figure 9.1b). The copulatory mounting lasts only some tens of seconds. After the male has dismounted, he usually continues to tend the female and mounts her again, perhaps in 10 minutes or so, but the interval and the number of mountings appears to vary with the male. From then onward, the pattern depends on whether the tending male becomes sexually depleted first or estrus terminates. In the former case, other males will tend the female.

Distribution of reproductive behavior

The annual distribution of the breeding functions is uneven. It appears that most, if not all, *S. c. caffer* populations drop some calves throughout the year. However, there is more than a suggestion of pronounced breeding peaks, which vary in time and intensity with the locality. Zuckerman (1953) gives a monthly distribution of 14 births in the London Zoological Gardens between 1872 and 1935. These occurred in February, March, May, June, July, August, October, November, and December. I was given at London's Regent's Park Zoo, in 1976, the birth dates of 9 buffaloes born at Whipsnade or Regent's Park between 1940 and 1975. These occurred in all the months from March to August, with the greatest number (3) in June, which was also the peak month for the 14 births cited by Zuckerman (it is interesting to note,

a

b

Figure 9.1(a) Precopulatory behavior: A male "chin-resting" an estrous female. (b) Copulatory behavior: A male mounts an estrous female, as two other males watch in tensed postures. The mounting lasts only some seconds. After dismounting, the male usually continues to tend the female and mounts her again. The interval and number of mountings varies with the male.

incidentally, that of these 9 births only 1 was male). Two animals of the *aequinoctialis* variety were born at the Antwerpen Zoo, Belgium, in April 1961 and May 1964. The births of these 25 animals cover all the months except January and September.

Ansell (1960) states that Northern Rhodesian (Zambian) records suggest calving throughout the year and gives authors (Verheyen, 1951; Verschuren, 1958; Jeanin, 1936) who arrived at a similar conclusion for the Garamba and Upemba National Parks, Belgian Congo (Zaire), and the Cameroons. However, he mentions the possibility of a peak in June–July in the Luangwa Valley, eastern Zambia, April to June in the Upemba area of Zaire, near the Sudan border (Verheyen, 1951), and the end of the dry season in the Cameroons (Jeannin, 1936), I presume about February. Pienaar (1969a, p. 39) states: "Contrary to earlier beliefs, it has now been established that buffalo in the Lowveld of the Transvaal have a definite and clearly demarcated breeding season. Although out of season births are not uncommon, it is now evident that the majority of buffalo calves in the Kruger Park are dropped during the period January to April, with a peak in January and February (Fairall 1967)." For the western Serengeti area, Tanzania, Sinclair (1977) shows a high peak of births from March to May, culminating in April, and a great drop in chin-resting, mounting, or copulation during August and September, while the peak for these activities is shown in (May)–June–July. Pienaar (1969a) states that in the Kruger Park, South Africa, a rutting season occurs from March to May and mating activity peters out during the winter months, I assume July to September, running into October. The only occasion when I did not record any young calves within an observed buffalo population was October 1972 in the Mabalauta area, Zimbabwe.

The gestation period of *S. c. caffer* is between 330 and 346 days (Vidler et al., 1963), and thus the calves will be dropped about 11 to 11½ months after conception.

Calving

My indirect observations suggest that most births take place in the small hours of the morning, but some occur at other times as well. On three occasions I have personally found or had reported to me a placental afterbirth by a calving site, indicating that it is not, or not always, eaten by the mother. I was able to fully observe only one calving, in April 1973 between 0806 and 1200 hours (the time when I stopped taking continuous notes). In the following transcript of field notes I assign a zero value to 0806 hours and express times in minutes starting from this base. The calving took place at the edge of the Busanga flats, north of the Kasolo Kampinga thicket, Kafue National

Park, Zambia. I estimated the female at about 5 years, after spotting a discharge from her anal/genital area:

Time: 000 Cow grunting repeatedly while walking; nervous, restless: swinging her head from side to side, halting with her head hanging down, stopping at the alert for no apparent reason; walking, then stopping and spinning around with her head near the ground, touching the ground with the tongue, then lifting her head and looking forward, then wheeling back to the direction of herd movement and walking on.

Time: 082 Cow is on left-rear of herd, very slow, dropping behind. Am remaining with her.

Time: 090 Cow is in a grassy patch near a tree, lies down, grunting occasionally. The herd is grazing and resting ahead, semistationary.

Time: 092 Cow is up, displaying grazing behavior but not actually grazing?

Time: 095 Cow wandered a little, a few meters, changing direction, and lay down again.

Time: 096 Cow is up and standing, a few grunts, looking ahead, then to the sides, switching the tail at frequent intervals.

Time: 098 As before. Moving a pace or two, grunting. At this time, she lowered her head and rubbed her forehead on the ground, both side to side and front to back.

Time: 101 Cow lies down, then gets up immediately, turns part-circle and lies down again.

Time: 105 Cow up, just standing, as though ruminating.

Time: 108 Cow lies down with grunt.

Time: 109 Cow is lying down, lifting her head slightly now and then, making jaw movements as if ruminating but more pronounced, with the sound of teeth grinding and impacting on one another.

Time: 113 As before. Occasional grunts and hitting the head on the ground.

Time: 115 Cow gets up with a grunt. Something is protruding from the vulva, hanging limply: membrane. Cow lies down almost immediately.

Time: 126 Cow has been lying down, now quiet, now thrashing about slightly and grunting. Now calf's feet can be seen.

Time: 139 Cow lying down as before. Quite a bit more of the calf is visible (legs).

Time: 157 As before until now. Now the cow moved as if trying to stand up, then relaxed for a moment, then tensed her body, lifting her head and hitting the ground with it. Plenty of fluid. Now head of calf can be seen.

Time: 159 Cow's body, including legs, jerking slightly as if by involuntary spasms – "rippling" effect. A little more of the calf is visible – not much more.

Time: 164 Cow tensed, relaxed, dropping her head on the ground, then tensed again, her head moving backward and slightly up (off the ground); she let out a vocalization much like that of a dying buffalo that has been shot. Moments later all of the calf is out.

Time: 167 Cow is down for a moment (less than a minute?), then she gets up clumsily, turns slowly and starts licking amniotic fluid near the calf. She stands, rump toward me, tail switching, her body blocking the calf. Cow's rump is wet.

Time: 170 Calf makes the first sound (high-pitched bellow?), moves on the ground. Cow is over the calf.
Time: 178 Calf is up, back humped and legs spread, shaking. Cow is standing.
Time: 181 Calf moves, falls forward, tries to get up, collapes, tries again, succeeds: moves forward. About 15 centimeters of umbilical cord hanging (remainder of umbilical cord had presumably been eaten by mother during licking).
Time: 183 Cow is moving slowly, grazing? – but more or less in one place.
The wind started to shift and a few buffaloes were coming, and I moved back about 120 meters into a clump of vegetation, to prevent being noticed (later I verified that the preceding observations were made from about 11 meters.)
Time: 204 Cow seems to be grazing more or less in place, the calf next to her.
Time: 213 Four cows and a bull, and two bulls somewhat apart, slowly approached, grazing, toward the place where the mother is. When quite near, one cow stands at the alert, facing the mother and calf, then resumes grazing.
Time: 229 All eight buffaloes plus calf are moving in scattered formation and very slowly, grazing and stopping, toward the rest of the herd, now some 700 meters away.
Time: 234 As above. Total times: 3 hours 54 minutes.

From conversation with the staff of the Antwerpen Zoo, in 1976, I gathered that, in one case, 2 days elapsed from the appearance of amniotic fluid to birth and that birth took some 10 minutes. Regarding the calving observed by me, I am unable to comment on how representative it is, since I did not observe any other, except very fragmentally.

Barren sexual behavior

Sexual behavior, that is, mounting and chin-resting, occurs between individuals that cannot conceive: females may mount estrous females, and the opposite may take place. Immature individuals mount one another during play. Adult males occasionally chin-rest other males and mount them. These interactions do not stimulate third-party interest, with the occasional exception of an estrous female arousing the interest of other females. Females with pronounced male secondary characters have been seen to participate in these interactions but not very greatly.

A classical recurrent example of a displacement activity in *S. c. caffer* is the mounting, by males, of females and other males during alarms and early stages of flight. This also occurs at other moments of excitement: I observed it in Zimbabwe and Tanzania in animals nearing water in hot weather, after many hours without drinking.

II. Mother–calf interactions and nonadult behavior

Mother–calf interactions of *S. c. caffer* during the early postpartum phase largely resemble those of domestic bovids of the genera *Bos,*

Ovis, and *Bubalus*. During the first day following birth the calf enters into a fundamental relationship with its mother, acquires suckling and basic locomotive ability, and may put its muzzle in contact with vegetation, although I did not observe nibbling at this stage. Throughout the first day the mother is very protective toward the newborn and does not normally leave it on its own. Identification of the calf by the mother appears to become established at that time and, probably to a lesser degree, identification of the mother by the calf. I base this statement on the observation that discriminatory behavior is shown more by mothers than by calves. Thus a lost calf may attract the attention of more than one mother searching for her young, but those mothers that fail to identify a calf as their own immediately lose interest and walk away. A lost calf, however, may initially follow a female that is not its mother. When the female fails to solicit or permit mutual proximity with the calf and, above all, discourages any attempt by the calf to suckle her, the calf wanders away, often vocalizing until contact with its mother is reestablished. I have not ascertained whether misidentifications occur.

During detailed observation of a calving (this chapter) the mother groomed the calf twice following parturition: For less than 1 minute within 4 minutes after giving birth, and for some 8 minutes starting about 6 minutes after giving birth. I saw other females groom calves that I judged to be several hours old. *S. c. caffer* mothers use the tongue less when grooming than domestic *Bos* mothers.

Newborn calves begin to show hunger before their coats lose the signs of recent birth, within a few hours after parturition, by exploring the mother's body with muzzle and tongue. As the calf presses and nudges the mother's flank and legs with its muzzle, more or less at random, the mother gradually walks forward until the calf is left slightly behind, then halts and allows the calf to catch up with her rump. The calf becomes increasingly more aggressive and exploratory in the pursuit of its mother's rump, until it finds a teat. This method of initial presentation of the mammae may explain why *S. c. caffer* young tend to suckle from the rear, between the mother's hind legs, while domestic calves of the genus *Bos* suckle mainly from the side, that is, the attitude in which they initially discover the mother's teat (Hafez, 1975).

Calves learn to walk during the first day following birth. Although newborn buffaloes are not as agile as some antelope fawns, for example, *Connochaetes* sp., my observations indicate that they may nonetheless walk effectively very soon after birth. In the calving described in this chapter, the calf stood for the first time 15 minutes after birth and started to walk 1 hour after birth. In another case, in the Busanga area during July 1974 I saw a female with a calf that I estimated as only hours old. The place was a recently vacated herd resting site, and the

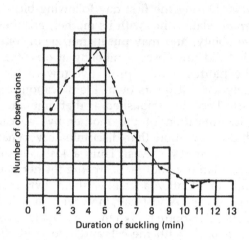

Figure 9.2 Observed frequency of suckling bouts of different duration in buffalo calves from various localities. The broken line is a three-item moving
average of the histogram. One may speculate that the irregularity in the histogram, i.e., a high incidence of suckling of 1–2-min duration, a very low incidence of suckling of 2–3 min duration and a high incidence of suckling of 3–
5-min duration, may have significance: the 1–2-min "high" could be caused by
aberrant suckling, whereas the "high" at 3–5 min could be caused by successful suckling. Thus the graph may be composed of two populations: one representing aberrant and the other successful suckling. The three-item moving-
average line, obliterates the depression between the two "highs."

time 0825 hours. I caught up with the herd and, at 0950 hours, that is,
85 minutes later, I identified the same female and calf at the end of the
herd. I estimate that the mother and calf had to walk at an average rate
of not less than 3.9 kilometers per hour, since the distance covered was
about 5½ kilometers.

Calves remain very close to mothers until some 2 months old. During this period grazing and drinking behavior, defensive aggression,
and play appear.

Sinclair (1977, p. 107) reports that calves suckle for periods of 3 to 10
minutes "at intervals throughout the day," which coincides with my
observations. In four sets of observations covering the entire 24-hour
cycle and two covering parts of the cycle, 126 hours in all, calves
suckled on an average 10.5 times a day ($n = 55$), averaging 3.4 minutes
per suckling bout. The data on which these figures are based appear in
Table 9.1. Of the five calves reported in this table, the first was observed during two 24-hour periods, 1 week apart (Lolgorien, Kenya);
there is little correspondence between individual suckling bouts during
the two periods. The oldest calf was number 5, from Mabalauta, Zimbabwe, estimated at about 4 months old, and none of the calves were
completely newborn. Figure 9.2 shows the frequency of suckling bouts

Table 9.1. Suckling bouts observed in S. caffer calves

Clock time	Duration of suckling (min)					
	1a	1b	2	3	4	5
0000–0100	0	0	8	*	0	0
0100–0200	0	0	0	*	0	0
0200–0300	0	4	0	*	0	3
0300–0400	0	0	5	*	3 + 5	0
0400–0500	2	9	2	12	0	0
0500–0600	5	0	0	0	4.5	1 + 7
0600–0700	0	0	0	0	0	0
0700–0800	0	1	5.5	1.5	0	0
0800–0900	3.5	0	*	1	6	5
0900–1000	0	6	*	0	0	0
1000–1100	9	2 + 1	*	4.5	4 + 4	0
1100–1200	0	0	*	0	0	5
1200–1300	0	5	*	6 + 2	0	0
1300–1400	7	0	*	0	4	0
1400–1500	0	3.5	*	4.5	0	13
1500–1600	0	2	*	0	7.5	0
1600–1700	6.5	0	*	9	0	0
1700–1800	0	0	*	0	0	4
1800–1900	0	1.5	0	4.5	4	0
1900–2000	0	0	2	0	0	0
2000–2100	4	4.5	0	*	0	0
2100–2200	0	0	0	*	6	6
2200–2300	0	11	4 + 1	*	0	7
2300–2400	2	0	0	*	0	0
Total minutes	39	50.5			48	51
Total bouts	8	12			10	9

Note: Column heading numbers represent one calf from Lolgorien, Kenya (1a, 1b), three calves from western Zambia (2, 3, 4), and one calf from Mabalauta, Zimbabwe (5). * indicates no observations were attempted (and thus no totals are shown in two columns). Duration of suckling within 0.5-min accuracy.

of different duration. This graph may be interpreted as showing two populations: A main one with a normal distribution and mean of some 5½ minutes' duration and a second one with a peak around 2 minutes duration. The former population may represent successful suckling bouts, the latter aberrant suckling. This would be compatible with six observations (Table 9.1) that very short suckling bouts often occur in rapid succession with one another. When the frequency of observed sucking bouts is plotted against the time of their occurrence the resultant distribution does not suggest any particular daily pattern. This is in agreement with suckling behavior reported for calves of domestic cattle (Hafez, 1975), in which no definite patterns were identified.

Figure 9.3. Calves are good runners in emergencies, capable of maintaining top speeds for a kilometer. Because of its smaller size, however the calf must run more intensively than adults to keep up.

In the first days or weeks following birth, depending on the individual, calves learn to run and to maintain a running pace for as much as a kilometer. During stampedes, calves are compelled to maintain their top speeds to keep up with the adults (Figure 9.3). Even then, they slowly lose ground, and hence contact with their mothers. As the herd slows down, calves catch up with their mothers by continuing to run, or by walking, directly ahead. Since buffalo herds tend to run in straight lines, this usually reunites calves and mothers. If a calf is unable to reestablish contact with its mother by this behavior, it starts to vocalize, is answered by one or more mothers, and makes its way to each in turn until it locates its own. A small calf searching vocally for its mother may be joined and closely followed by an older nonadult or adult of either sex, until the mother is located (Mloszewski, 1974). Such an animal need not be a member of the calf's basic herd.

It is reported by Sinclair (1977) from the Lake Manyara National Park, Tanzania that the playing of a tape-recorded calf distress call had the effect of repeatedly attracting an entire buffalo herd toward the vocalization. On the several occasions that I imitated a calf distress call, in the Busanga and Chunga areas of western Zambia, Ruaha National Park, southern Tanzania, and the Mabalauta area, southeastern Zimbabwe, some of the adult buffaloes within my sight displayed interest and alertness, and one or more females eventually answered the call. Sometimes I was able to lure these females to within 5 meters of me.

Once a female's interest was aroused, almost every one of my calls was answered with a single reply. In one case, in the Ruaha National Park, several adults of both sexes approached my imitation calf call to within some 10 to 15 meters, then lost interest, with the exception of one female which continued to act interested and excited. Then, as she was joined by a large calf, possibly her own, she too lost interest. Such evidence implies a dual response to calf distress signals: One response is by any of the herd members and may be collective, which, as far as I know, has never been actually observed to go beyond exploratory behavior; and another response, by individual females, apparently searching for their own calves. In the latter cases, it remains to be ascertained whether a fortuitous substitute calf is ever acceptable.

The buffalo mother is solicitous of her young calf when it is near her but her concern appears to rapidly diminish when the calf becomes lost. The mother displays little restlessness, apart from making an initial effort to locate the calf by following and answering what may be its vocalization, but only for short distances away from the herd. Of the six calves left behind moving herds that I observed, none were rejoined by their mothers during 1 hour or more of waiting. In agreement with my own experience, Ludbrook (1963) reports from the Kafue National Park the abandonment by a mother of a young calf estimated as 1 day old, when it became stuck in mud. The mother followed the herd and failed to return within 45 minutes. Again, Yamba (1969) reports waiting for several hours near a female calf, estimated as 2 days old, lost in the Busanga area, to see if the mother would return, with negative results. An instance suggesting that exceptions occur is reported from Uganda (Uganda Game Annu. Rept., 1934), when a female fled, presumably at the approach of a game guard, then returned to look for her calf, knocking the guard down in the process but without pressing a serious attack.

A lost calf of several days or older may appear to know the direction in which its herd has gone: If one attempts to lead such a calf in a different direction, it may resist vigorously, although it does not object to being led in the direction taken by the herd (Mloszewski, 1974), nor does it struggle when it is being carried in any direction.

Calves of about 2 months occasionally graze for minutes without interruption and are able to cope with sparse dry season pasture (e.g., Figure 2.7a). They still suckle frequently, but it is at this stage that the amount of suckling begins to vary rather considerably between individuals and according to the season, that is, probably according to the amount of milk produced by the mother. Calves of this age go through the paces of drinking with the adults but do not appear to take in much water, which is understandable since they are still largely on a milk diet. Since they stay close to their mothers, there is little association among calves of this age.

Figure 9.4. Buffaloes appear not to recognize objects visually at short ranges: This female and juvenile, lacking auditory or olfactory clues, looked straight at me, then grazed past, as I sat on the ground almost motionless. The female was herding several nonadults. (Southeastern Zimbabwe.)

Calves that are weaned early, that is, between 4 and 6 months of age, because of loss or noncooperation of the mother, tend to associate with one another if they meet. They try, moreover, to follow adult females, which may respond aggressively or evasively. If such calves attach themselves to an adult female and are not rejected, they remain by her side. In my experience, this female is always old or in advanced maturity. The continued proximity of the calves appears to elicit some quasi-maternal responses in such females, and they may begin to herd the calves. In heavily hunted or very large buffalo herds, where the probability of orphans may be high, several calves may congregate behind one female, and the group may be further augmented by older juveniles. Recruitment into such a juvenile club may be sufficiently continuous to keep it in existence for as long as the "nanny" female is able to lead it. These females do not merely tolerate the calves but show concern for them, especially in matters of safety. In Lolgorien, Kenya an old female "in charge" of two to five calves always reconnoitered at length before leading the calves out to graze at the end of a midday rest, and was usually the first member of the herd to move toward pasture. Figure 9.4 depicts another such female, from Mabalauta, Zimbabwe. I regard this behavior as generated by a set of coincidences: If motherless calves, as well as an adult female of low discriminatory threshold, exist within a herd, then this pattern has a good

Figure 9.5 Imitative behavior: The old male looked up while grazing; the juvenile and the young adult male looked up simultaneously. This is a common type of juvenile and subadult behavior, possibly learning.

chance of appearing. Nonadults following a "nanny" place much trust in her and take all their cues from her; they stop and move as she does. It was my impression, when very close to such groups, that some calves spotted me while the "nanny" did not. In these cases the calves did not raise an alarm but adhered to the female. This dependence on adult judgment has also been observed by me in calves with mothers. In general, the behavior of nonadults accompanying adults is largely imitative (e.g., Figure 9.5).

Early play is sterotyped and consists of running in circles, prancing, and butting, usually the mother. Similar play occurs in many bovid species, and I observed it also in warthogs (*Phacochoerus aethiopicus*).

Reliance on milk appears to decrease from about 4 months of age onward, but the gradual weaning process may be considered to start even earlier since after reaching 2 months calves are often discouraged by the mothers from suckling at will. The rapidity of weaning varies considerably, depending on the mother's condition, the individual calf, and perhaps the season. Calves esimated as 9 months old may still attempt to suckle, while others, usually through necessity, do not get milk after 4 or 5 months of age. The behavior of calves of over 6 months differs from that of adults in the incompleteness of the weaning process and their reliance on adults in initiative taking, especially in emergencies. The greater skeletal suppleness, longer legs in relation to the trunk, and light horns permit the calf to curl up when lying on the ground and to scratch itself with a hind leg more easily than an adult. The greater leg length relative to body size may also help in overcoming running disadvantages caused by the calf's small size.

As a calf gains in confidence it leaves the mother's side more and more frequently until, between the ages of 1 and 2 years, the juvenile male leaves its mother altogether, to associate with other males, while the juvenile female assumes a subordinate adultlike role, often associated with the mother.

10

Other behaviors

This chapter deals with several behaviors that are not closely related and that do not fit well into any of the other chapters. Individually they can be discussed only briefly, and do not warrant separate chapters.

Individual movements and self-grooming

When an African buffalo encounters small obstacles in its path, such as remnants of small termitaries, dead wood, small shrubs, or cobbles, it usually goes around them, even in cases when it may appear easier to go over them. It is reluctant to walk fast, possibly because the posture that it must then assume (Chapter 5) may hinder digestive processes and activities, and it does not normally run except in emergencies. When it does run, however, the protection given by its very thick skin and heavy body permit it to move in nearly straight lines through dense bush, bypassing only trees. Although its movements are usually slow, the African buffalo has considerable agility and body control, as shown in Figure 10.1. The animal photographed was lying down an instant before the photograph was taken, became alarmed, possibly by a glint on my camera lens, and attained the posture recorded, with the anterior parts of the body off the ground and balanced on the haunches, in one continuous movement. This is notwithstanding the frequent observation that the normal way for a buffalo to rise is by first standing on its hind legs, while the front remain kneeling, and only then to stand on the front legs, all of which takes some 3 to 5 seconds and is a behavior common among bovids. In areas where large termite mounds occur, buffaloes do not hesitate to climb their steep gradients to browse; and occasionally, convergently with bubal hartebeests (*Alcelaphus buselaphus*), single animals climb steep high places to scan the surroundings.

Figure 10.1. This individual demonstrates the occasional great agility of *S. c. caffer*. It is in the early stages of rising from a recumbent position, on becoming alarmed. The strength required to lift the body in this way must be considerable. The animal was lying down, became alarmed, possibly by a glint on my camera, and attained the above posture in one continuous movement. This is not the normal way for the buffalo to rise (see text).

Despite the relative shortness of the African buffalo's legs, it is capable of scratching some of its anterior parts, including the ear and jowl, with the hoof of a raised hind leg, in the manner of lighter-built species, such as domestic cattle or wildebeests.

When to relieve itches buffaloes scratch their bodies against trees or termite mounds, they stand with hind legs close together, front legs spread wide, seesawing back and forth while leaning sideways against the objects. This is a moderately frequent ordinary behavior. Buffaloes rub or scratch the fronts, tops, and backs of their heads, as best they can, against trees and on the ground. This also is a moderately frequent ordinary behavior, observed as well in other bovids, for example, eland (*Taurotragus oryx*). An escalation of this behavior is described in Chapter 7.

Ear, tail, and skin movements

Ear twitching and tail switching tend to accompany each other and are common behaviors in both subspecies of *Syncerus*. Ear twitching is not reduced on lying down, but tail movements become fewer or cease. Resting buffaloes, in the presence of flies, twitch their ears continuously at observed rates of one twitch about every 1.7 seconds. Continuous regular tail switching was observed to occur most often at the rate of one back-and-forth excursion about every 2.6 seconds, which is less than twice the rate of rapid continuous ear twitching. Hence, there is no obvious correspondence between each one-way swing of the tail and one movement of the ears. Clearly, each ear movement is not necessarily accompanied by a tail movement, and the reverse is also true. Tail switching may attain a rate of one back-and-forth excursion in 1.7 seconds, but this does not persist and appears to occur only under stress. Ear twitching is reduced or stopped during deep rest. Slow rhythmic or irregular tail switching usually accompanies grazing. Both ear and tail movements may become sporadic and irregular.

The rates of ear and tail movement in *S. c. caffer* appear largely functional. Thus when the numbers of flies increase, so do ear and tail movements, which serve to discourage flies from landing on some thinner-skinned parts of the buffalo's body, that is, portions of the head and the anal–genital zones. Ear and tail movement is more frequent during the wet than the dry season. I correlate this with greater numbers and wider distribution of flies in the buffalo's wet-season habitat. An 88 percent coincidence was observed ($n = 116$) between an increase or initiation of ear twitching and gusts of wind. A possible explanation is that flies sitting on the buffalo are likely to be blown away from the host if dislodged at a moment of fast air movement. The correlation between ear twitching and wind gusts appeared also to exist in two captive buffaloes of the *caffer* subspecies, although it was based on very small samples ($n = 6$ and $n = 13$). Regular continuous ear and tail movements were observed in captive individuals of the *caffer* subspecies, although they were seldom closely spaced. Occasional ear and tail movements are common in captives.

I have made a small number of observations which suggest that ear twitching is more regular in the *nanus* subspecies. Three Ghana bushcows (*S. c. nanus* var. *brachyceros*) and a similar captive adult male of uncertain provenance all twitched ears about once every 5 seconds, when resting or stationary, with considerable regularity. The ear movement was usually, but definitely not always, accompanied by a brisk tail switch. It remains to be confirmed whether ear and tail movements are reflexive, blinklike, or deep-rooted habitual behaviors. I am inclined to view them as the last.

Spasmodic skin shivers, similar to those in thinner-skinned species such as horses, have been occasionally observed in *S. c. caffer*, but are uncommon.

Allelomimetic behavior

The most frequently observed allelomimetic behaviors, that is, nearly simultaneous and mutually stimulated performance of one activity by all or most group members, included assumption of the alert stance, flight, and resumption of grazing at the end of a resting period. Simultaneous assumption of the alert stance occurred when a grazing group sensed danger at night. I observed it on several occasions by moonlight and suspected it, on auditory evidence, during moonless nights when visibility was inadequate. In so far as I was able to observe, all the individuals halted and raised their heads nearly together, remaining motionless for as long as a minute, then resumed grazing one by one or moved away. I was unable to detect any signal that might initiate the behavior, which was positively observed in groups of around 100 individuals with an average spacing of not less than 2 meters. I observed similar behavior in a small group of bushcows (*S. c. nanus*) in daylight in Ghana during 1962.

Simultaneous flight was always preceded by a state of alertness on sensing a threat, of at least 5 seconds but usually longer, which conditioned the buffaloes for the eventuality of flight. At the instant at which flight was begun by all the group members, they may have stood in a wide-front array 20 or 70 individuals across, facing the threat. Without an observed signal, all the individuals turned right or left together and began flight. If there was a flight signal it must have been passed across the group's wide front, as well as front to rear, resulting in a nearly instantaneous general and coordinated flight response.

The observed "nearly simultaneous" resumption of grazing at the end of a resting period starts less in unison than the previous two behaviors and fits the definition of allelomimetic behavior only marginally. The stimulus involved the classical concept of social facilitation: A high-status individual, usually the top male, became active – for example, walked several meters from its resting spot – and started to graze. Between 13 and 60 seconds thereafter, the remainder of the group, which may have by then abandoned their resting stances, moved in a body and lowered heads to start intensive grazing. This behavior was not accompanied by vocalizations usual at the end of a resting period, which I interpreted as signals to resume activity (Chapter 11). The two behaviors appeared to be alternatives, the silent "allelomimetic" one most common in heavily hunted (and therefore in practice relatively small) groups, living under a high state of stress. But most often group alertness, flight, and beginning of grazing did not occur with a spontaniety suggestive of allelomimetic response and spread gradually.

Within basic herds and herds, there is an obvious tendency to follow and associate. This tendency does not attain the high level observable in domestic sheep (*Ovis aries*) but appears strong nevertheless. The order in which individuals are associated at a given moment, however, is by no means rigid. Tulloch (1978, p. 334) commented when he discussed close associations in the Australian feral water buffalo (*Bubalus bubalis*) while on the move: "This does not mean that each individual always remained next to the same ones, but rather that all buffaloes in a group behaved in the same way. This happened on every day throughout the whole study period." Thus it appears that *Syncerus* and *Bubalus* behavior may be convergent in this respect. This behavior has more functional application to the Australian feral water buffalo, since the latter's daily regimen (Tulloch, 1978) is more regular than that of *Syncerus*, and therefore able better to accommodate, and perhaps benefit from, coordinated routine patterns. It is reasonable to assume that an allelomimetic response related to a routine behavior is an advantage to groups whose individuals never or very seldom function separately, and therefore do not depend on, and need not cultivate, individual initiative. If strong allelomimetic tendencies inhibit single initiative, which appears likely, the value of allelomimetic behavior drops as situations in which individuals must act independently of the group increase.

A tendency to defecate together was occasionally observed in buffalo herds, both stationary (e.g., Busanga, Zambia) or walking (e.g., Mabalauta, Zimbabwe), and I had the impression that social facilitation was operative. Parsons (1966) speculated that defecation in domestic cows may have sometimes an allelomimetic character.

Other-interested behavior

Described in Chapter 9 were two types of other-interested behavior related to calves: The caretaking of calves by females other than their mothers, and "escorting" (trailing) of lost calves by nonadults or young adults until the calves found their mothers. The first behavior clearly belongs in the class of "other-interested"; the second, however, is uncertain, since it may be explained without introducing a desire or drive to protect the lost calf: Perhaps a lost calf's search behavior can stimulate similar, aberrant behavior in an older individual if the latter retains a memory of its calf stage, and thus the act of following the lost calf may not have any caretaking connotation.

Other-interested behavior occurs between old bachelor males whose faculties are partly impaired. I observed in Lolgorien, 1965, a pair of old males, one of which was blind and probably deaf, while the other was much less affected by old age. Before walking off, the latter bull always rubbed against the impaired one, and this appeared to serve as a signal to move. When it was time to stop, the healthier bull blocked the other's way, that is, let the impaired bull walk into him; then both grazed, drank, or rested. In emergencies, if the impaired bull sensed them, he stood at the alert waiting for the other to give him a direction of flight by bodily contact. If only the healthier bull sensed danger, he went out of his way to signal his partner by means of body contact before fleeing. The tactile message appeared to include both information of impending threat and flight direction. Although this partnership was particularly well developed and strongly bonded, I have observed elements of similar behavior in other associations of *S. c. caffer*, both male and female.

I observed an extreme instance of other-interested behavior in October 1970 in the northern sector of the Kafue National Park, as I was following a part of the Ntemwa herd. There was an adult male with a swollen lower jaw (caused by infection or snakebite?) and acting sick. He lay down frequently and was losing ground with respect to the moving herd. Two subadult males remained with him. When on two occasions the sick bull was in danger of being left behind, the subadults nudged and generally encouraged him by bodily contact to get up, apparently for the purpose of catching up with the herd. The subadults remained near the sick animal throughout my 8-hour contact with the herd.

Attack by lion against one individual may provoke aggression by others. I witnessed this behavior near the Mara Bridge, Kenya in open high grass, at about 1710 hours in August 1965. Two large nomadic male lions attacked one of three bachelor buffalo bulls and were in turn charged by the others. The lions withdrew, abandoning the wounded buffalo whose scapular region was lacerated. The three bulls, after an instant at the alert, started to walk away, the walk soon changing into a rocking lope, until they disappeared from view. I had lost sight of the lions but 8 minutes after the buffaloes disappeared I saw the lions' manes, one close behind the other, above the grass, moving at a walking speed in the same direction as the buffaloes. Similar behavior has been reported from elsewhere in East Africa (e.g., Cullen, 1969; Schaller, 1972, p. 260). There is here a possibility that the stimulant is not a desire or drive to protect a particular conspecific but rather to remove the lion threat in a more general sense, since buffalo aggression against lions does not occur only when a buffalo has actually been attacked; it may also take place simply in response to lion presence (e.g., Chapters 7 and 8; Cullen, 1969; Schaller, 1972).

11

Vocalization

Vocalization was studied by direct listening, from recorded tapes, and from sonograms. The tape-recordings were produced during 1977 and early 1978, when two Philips EL 3302 B/76P casette recorders became part of my field equipment. Some of the tape-recorded material was later converted into sonograms. A shortcoming of the direct listening method was the impossibility of exactly reproducing in writing the quality of a particular vocalization. After becoming familiar with the main sounds voiced by buffaloes, I initially named them by association with some other sound, as for example the *creaking gate* call, with qualifiers such as *long, short, low-*, or *high-pitched*, and so forth, and later such names were reduced to a letter code, also supplemented by qualifying words. This "shorthand" could be used only for well-known vocalizations, whereas less familiar sounds had to be described in field notes at some length. Since a written description of a sound is, at best, only partly explicit, the method was a long way from perfect in its capacity to impart my knowledge of a sound to others. The quantitative aspects of vocalization, that is, its presence or absence, its frequency, intensity, and distribution in time, were much easier to put down in writing than the qualitative aspects.

Although tape-recording of vocalizations did not simplify their written explanations, it did furnish standards to which reference could be made. It also made repetitions of exactly the same vocalization sequence possible. The disadvantages of tape-recording were those arising from the need to carry and operate sensitive fragile instruments under trying field conditions. I found that anything but a quite small model of tape recorder was out of the question for the type of study in which I was engaged, because of unwieldiness and extra weight. The price of carrying more sophisticated equipment would be the frequent loss of opportunity to record, as well as loss of all-round mobility.

201

Buffalo vocalizations appeared sufficiently complex to merit some kind of subdivision for study purposes. Some vocalizations within a buffalo herd appeared to have a collective significance, while others could not be correlated with any group activity and were related to circumstances not involving the herd as a whole. It was found convenient, therefore, to consider buffalo vocalization under two pairs of aspects: The amount as contrasted with the kind of vocalization, and group as contrasted with individual vocalization.

Group vocalization

The quantitative aspects of buffalo group vocalization were observed in Zambia, Zimbabwe, and Kenya, which furnished 59, 24, and 17 percent of the raw data, respectively. More specifically, the study localities were the northern sector of Zambia's Kafue National Park, mainly the Busanga area; the Mabalauta area of the Gona-Re-Zhou Game Reserve in southeastern Zimbabwe; and the Lolgorien area of southwestern Kenya, situated between the Mara and Migori rivers.

The absolute amount of vocalization varied of course with the number of individuals in a herd. I was interested mainly, however, in the relative amounts of group vocalization at different times, that is, changes in its levels. I was able to observe that most vocalization in buffalo herds is not random but associated with activity patterns. Because activity tends to be uniform within a herd at any one time, the transitions from one vocalization level to another are rather definite, and relatively little affected by herd size. Such differences as must occur due to herd size can be taken care of by assigning rather wide ranges to intermediate levels of vocalization. The fact that the intermediate vocalization level is expressed as a range that takes into account the variation in herd size may cause the boundaries between vocalization levels to be somewhat less sharp than they would be if it were narrowed down to express a constant herd size, but it does not obliterate those boundaries, and the main group vocalization pattern is still shown. A human analogy exists in theater audiences. The size of audiences varies, though between reasonably well established limits. The larger the audience, the more chatter can be expected during the performance. But this during-performance chatter normally does not exceed a certain low value, and does not normally mask the abrupt change in chatter level at the start or finish of an intermission.

In noting group vocalization, a "moderate" level was defined as three to six individual calls per minute, during at least 5 minutes; anything less was defined as a "low" level of vocalization; and a greater-than-"moderate" vocalization level was defined as "high." The result of group vocalization was usually the production of a certain amount of sound, rather than the participation of all or most individuals present in making the sound. This is in contrast with another type of collective vocalization, encountered in domestic cattle of the *taurus* species, especially in conditions of congestion and high stress (e.g., some slaughter yards), in which every individual attempts to participate, resulting in a subhomogeneous continuity of sound, and which was encountered but seldom in African buffalo herds.

Periods of no vocalization at all were also noted. They occurred even in very large buffalo groups, that is, exceeding 1000, and their duration was comparable to the duration of silent periods in small herds of as few as 30 individuals. This suggested that certain types of group vocalization were not merely an aggregate of unrelated individual calls. If the latter were the case, one would expect much more continuous vocalization in big herds than in small ones. These types of vocalization appeared, therefore, to have a strong social or group basis, largely independent of group size.

Group vocalization was not found to vary much from locality to locality, or from one end of the year to the other, so long as the intensity of seasonal climatic features such as rain, drought, or wind remained moderate. Marked intensification of these seasonal climatic features did bring about short-lived changes in vocalization patterns, which, however, were not statistically prominent. Figure 11.1 compares the occurrence of low to nil vocalization with the occurrence of moderate to high vocalization over the 24-hour cycle. The frequencies of each of the two classes of vocalization are plotted as percentages, against a given time of day. Dry- and wet-season vocalization and data collected in different localities are given on separate plots. The distribution of the two classes of vocalization over the 24-hour cycle is broadly similar throughout. Furthermore, the average relative amounts of the two vocalization classes vary little between the dry and the wet seasons, and no significant trend was detected. The low-to-nil vocalization class is invariably the more frequent, averaging 67 percent of all vocalization during the 24-hour cycle and ranging from about 60 to about 80 percent at any one moment. I have observed that the northern Tanzanian and Kenyan buffalo herds were less vocal than the more southerly ones. Figure 11.2a for Lolgorien, southwestern Kenya, exemplifies this situation since it shows the highest percentage of low to nil vocalization (79 percent).

Normally, group vocalization is initiated by a few individuals, at or near the top of the dominance order, or mothers, calves, and juveniles. This stimulates like behavior in some nearby animals and, to a lesser degree, in some more distant animals. Thus vocalizations of the group type originate from one or several "vocalization centers" within the herd, spreading to involve other individuals. The latter, however, usually number at any one time well below half of the entire herd. Only

204   *The African buffalo*

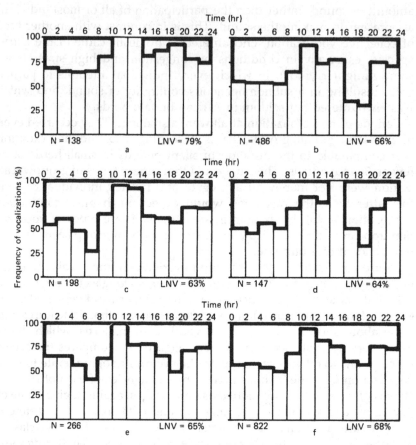

Figure 11.1 Group vocalization. Frequencies of vocalizations expressed as percentages of total vocalization (including nil vocalization) in each portion of the figure. The lower histograms represent low/nil volcalization (LNV), while their complements (heavy black internal borders) represent high/moderate vocalization. N = observed total. (a) Lolgorien, Kenya; (b) Busanga, Zambia; (c) Mabalauta, Zimbabwe; (d) wet season; (e) dry season; (f) composite.

very rarely do most of the individuals forming a herd vocalize simultaneously. The sound range of this latter type of collective vocalization is shown as a sonogram in Figure 11.2.

The quantitative or semiquantitative aspect of group vocalization is considered of interest irrespective of the quality of the individual calls involved. It does not appear necessary to differentiate between dry- and wet-season data for this purpose because, as already stated, no evidence was found of significant seasonal variation in group vocalization of *S. c. caffer*, when the raw data were analyzed by season and habitat. All data were assigned to one of four vocalization levels – high,

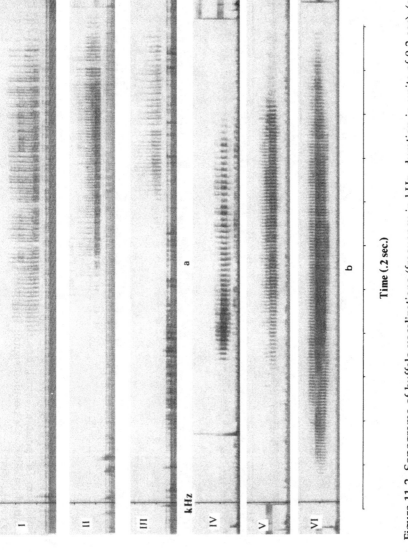

kHz

Time (.2 sec.)

Figure 11.2. Sonograms of buffalo vocalizations (frequency in kHz, duration in units of 0.2 sec): (a) Sonograms I, II, and III – signal to move, described in the text as vocalization A. (b) Sonograms IV, V, and VI – direction-giving signal, described in the text as vocalization B.

205

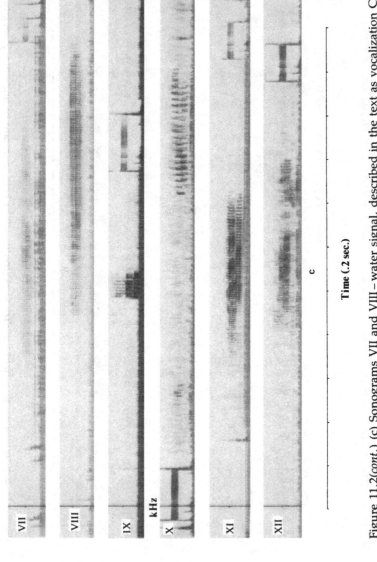

Time (.2 sec.)

c

Figure 11.2(*cont.*) (c) Sonograms VII and VIII – water signal, described in the text as vocalization C. Sonogram IX – position signal, described in the text as vocalization D. Sonogram X – a low croak, emitted possibly as a position signal. Sonograms XI and XII – warning signal, described in the text as vocalization E.

206

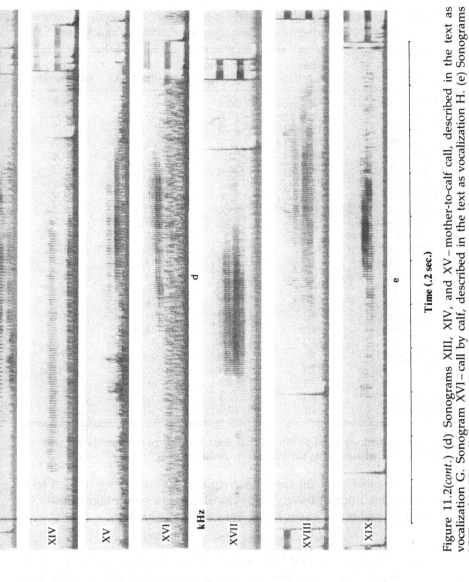

Time (.2 sec.)

Figure 11.2(*cont.*) (d) Sonograms XIII, XIV, and XV – mother-to-calf call, described in the text as vocalization G. Sonogram XVI – call by calf, described in the text as vocalization H. (e) Sonograms XVII and XVIII – calls by juveniles. Sonogram XIX – fragment of prolonged general vocalization by a herd of over 300, walking to water in the dark (0300–0400 hours).

The African buffalo

Table 11.1. *Average hourly percentage of four vocalization levels in S.* caffer, *based on 822 observations from Zambia, Zimbabwe, and Kenya*

Hour	High vocalization	Moderate vocalization	Low vocalization	Nil vocalization
0000–0100	32.4	26.5	32.4	8.8
0100–0200	11.1	14.8	37.0	37.0
0200–0300	11.4	34.3	37.1	17.1
0300–0400	18.5	22.2	55.6	3.7
0400–0500	8.7	21.7	39.1	30.4
0500–0600	25.0	34.4	28.1	12.5
0600–0700	37.9	24.1	31.0	6.9
0700–0800	25.0	15.6	37.5	21.9
0800–0900	8.3	19.4	63.9	8.3
0900–1000	3.2	35.5	38.7	22.6
1000–1100	3.1	0	65.6	31.3
1100–1200	0	9.5	59.6	31.0
1200–1300	7.7	11.5	53.8	26.9
1300–1400	0	18.2	30.3	51.5
1400–1500	0	16.2	59.5	24.3
1500–1600	15.2	17.4	45.7	21.7
1600–1700	9.1	18.2	38.6	34.1
1700–1800	7.0	27.9	44.2	20.9
1800–1900	18.4	44.7	34.2	2.6
1900–2000	7.5	17.5	45.0	30.0
2000–2100	23.1	15.4	42.3	19.2
2100–2200	6.1	9.1	66.7	18.2
2200–2300	7.1	11.9	57.1	23.8
2300–2400	20.6	17.6	38.2	23.5

Note: The sum of any line through all four percentage columns is 100 percent, but the actual sum may vary due to rounding.

moderate, low, and nil – and the frequency of occurrence of each level during each hour of the day was calculated as a percentage of the total vocalization for the given hour. Thus, for any one hour:

$$\%HV + \%MV + \%LV + \%NV = 100\%$$

where HV, MV, LV, and NV are high, moderate, low, and nil vocalization, respectively. The overall average figures obtained for the four levels of vocalization are shown in Table 11.1. When the nil vocalization values were examined in relation to the time of their occurrence it was noted that eight nil vocalization maxima of variable magnitude were present in the 24-hour cycle, at fairly even intervals. These maxima had values of 21.9 percent to 51.5 percent (average: 32.5 percent), compared to the overall nil vocalization average of 22.0 percent.

The best-established time of minimum vocalization, independent of season and locality, is between 1300 and 1400 hours. In that interval

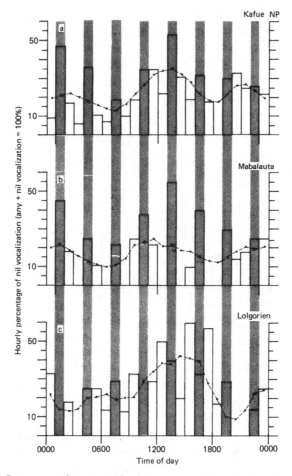

Figure 11.3. Group vocalization. The histograms show the average proportions (percentages) of each hour in the diurnal cycle during which vocalization was recorded absent. The shading underlines a tendency for silent periods to occur at 3-hr intervals. The coincidence is marked in a and b but much less so in c, perhaps partly at least because of the small sample size of the last. The broken lines are three-item moving-average series that eliminate the hourly fluctuations and show that in all three cases the occurrence of silence is most frequent between 0900 and 1800 hours. (See also Figure 11.4.)

the percentage of observed high vocalization is zero and nil vocalization 51.5 percent, while low-level vocalization exceeds moderate (30.3 and 18.2 percent, respectively). The halfway point between 1300 and 1400 hours was taken as fixed and from it 3-hour intervals were examined through the 24-hour cycle. The coincidence of these 3-hour intervals with nil-vocalization maxima is consistent. When the total data were plotted separately by the three component localities (Figure 11.3),

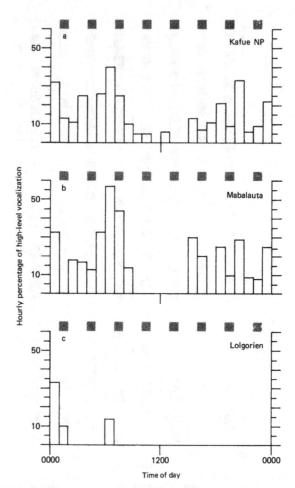

Figure 11.4. Group vocalization. This figure is complementary to Fig. 11.3. The histograms show the average proportions (percentages) of each hour in the diurnal cycle during which high vocalization was recorded. The shaded markers correspond to the shading in Fig. 11.3, which underlines the tendency for silent periods to occur at 3-hr intervals.

the Zambian (a) and the Zimbabwean (b) data fitted the 3-hour pattern, whereas the Kenyan (c) data, although showing a not incompatible pattern, were less regular. This could be at least partly ascribed to a rather low number of observations in the Kenyan sample (35) as compared to the other two samples (176 and 72).

High-vocalization maxima show a tendency to occur in the intervals between nil-vocalization maxima (Figure 11.4). On the Zambian and Zimbabwean locality graphs the preference of the high-vocalization "peak" to occur closer to one or the other of the two flanking nil-vocalization "peaks" is very slight. In 63 percent of cases it occurs closer to the nil-vocalization "peak" which follows it than to the one that precedes it. In the Kenya locality a high level of vocalization was a rarity, and for most of the graph the high-vocalization curve is at zero percent.

The significance of these fluctuations is uncertain. If they are real, the reason for such a rhythm may be physiological and related to the digestive cycle. For instance, cud-chewing periods in sheep number about eight in the 24-hour cycle (Fraser, 1974), and such periods of chewing are associated both in sheep and cattle with low levels of vocalization (Hafez, 1975). However, no striking correlation was noted so far between these fluctuations and rumination data for the African buffalo.

Types of individual vocalization

The range of buffalo vocalizations varies between approximately 0.1 and 2.3 kHz (1 kHz = 1000 cycles per second). Many of the common adult vocalizations resemble one another, but after repeated exposure I began to distinguish between them and to associate some of them with particular activities or events. I may have distinguished between vocalizations by ear alone, as I believed I did, but there were no objective means to ascertain the degree, if any, to which other factors, such as the time of day, were subconsciously used. From the buffalo's point of view, it is possible that circumstances under which some vocalizations are emitted constitute part of the signal.

Many buffalo vocalizations resemble the sounds commonly made by domestic cattle (*B. taurus*), but are generally displaced toward lower frequencies. As in domestic cattle, buffalo calves do not emit sounds of as low frequency as adults and subadults. However, the highest recorded frequencies are overtones of some adult vocalizations. The largest males tend to emit the lowest sound frequencies. Individual vocalizations, when produced at a high rate by a herd of buffaloes, are almost always loud. On the other hand, when the rate of vocalization is lower, the magnitude of the individual calls can vary over the complete range of volume from near-silence upward. The sound of buffalo vocalizations can be continuous and smooth, or "gritty"; extended over several seconds and tapered off at the end, or short and explosive. In a sample of 50 assorted single vocalizations, the duration varied from 0.3 to 4.1 seconds, with an average of 2.2 seconds. The spacing between vocalizations emitted by one individual is usually considerable and irregular, but may average as little as 3 seconds for several minutes at a time. Buffalo vocalizations appear to include communicative–cognitive signals, and are rarely emitted by isolated individuals. Some types of event are frequently or even normally accompanied by a corresponding vocalization, while others are only rarely accompanied thus. In my own transliteration, most buffalo vocalizations can be rendered approximately by "maaa. . . ," "ma," "mma," "bo," "bo-o-o . . . , " "bo-o-o," "nangh," "waaa . . . , " "wa," "mmm," "mmm . . . , " "nnn," "nnn . . . , " "aaa. . . ," and "aaa." Listed below are nine correlations between vocalizations and activities or events:

A. Signal to move, typically given at the end of a major resting period and in cases of sharply defined, often routine moves from one locality to another. This vocalization is usually emitted by a high-status animal, most often the top male. It is a low-pitched 2- to 4-second vocalization repeated several times at 3- to 6-second intervals. It is a coordinating signal, rather than an "order" to move, since it occurs often after a resting herd has begun to stir in place.

B. Direction-giving signal. This vocalization is continued intermittently by individuals in the lead for some time after a herd has been put in motion. The sound is "gritty" and monotonous, and originally prompted me to call it the "creaking gate vocalization." Individual vocalizations have been measured to last from 1.6 to 2.8 seconds.

C. Water signal, which precedes and accompanies movement toward water for the purpose of drinking. It is an extended "maaa . . . ," frequently terminated on an "inconclusive" note, with an expiratory sigh. When it is repetitive, it is emitted to a regular breathing rhythm, and in the latter half of an individual emission it may wander slightly up and down the scale, giving a moaning impression. It is emitted by one or a few individuals, at rates occasionally reaching 20 per minute, with individual emissions up to 2.5 seconds duration. The vocalization occurs when a herd has to walk a few hundred meters or more to water, but may be emitted even when movement is minimal in the wet season. This vocalization has been heard emitted a few times during the 20 minutes preceding actual movement to water, before becoming frequent and regular. Usually, however, water signals start much nearer to the time of movement toward water, that is, are responded to more rapidly.

D. Position signal, emitted by the higher-status individuals, apparently signifying: "I am here; do about it whatever my status demands." This type of vocalization has been measured at 2.7 seconds duration but it may be very brief (0.3 second).

E. Warning signal. A more intense form of D, used agonistically as a warning to an encroaching lower-status individual.

F. Aggressive signal. An explosive grunt that, when extended, becomes a sequence of grunts or a rumble. It is also used by high-status males at the time a stampede ends.

G. Mother-to-calf call, emitted by females in search of their calves.

H. Calf distress call, resembling G but not descending to as low frequencies.

I. A vocalization which may mean. "danger, lion." I heard it only three times, in western Zambia, in all cases during the day, when hunting lions were detected from far away by a member of a resting buffalo herd. It was an extraordinarily drawn out "waaaa" sound. Each time this vocalization occurred, the herd moved away at a purposeful walk and in compact formation.

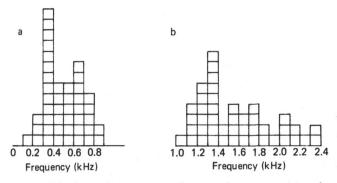

Figure 11.5 Individual vocalization. Distribution of minimum (a) and maximum (b) sound frequencies in 44 recordings of seven buffalo vocalization types.

A number of brief bellowings, grunts, honks, and croaks that I call "grazing vocalizations" is emitted while at pasture. These indicate a herd's position as it grazes forward, which may be useful to peripherally situated individuals, and also imply that all is well, but it seems unlikely that they are intended as specific signals. A number of vocal sounds made by buffaloes were not correlated with particular activities or events, to the exclusion of others. The death bellow of a mortally wounded buffalo, recurrent in hunting lore, is real and caused by a probably involuntary expulsion of air.

In a comparison of sonograms of several vocalizations that I felt able to distinguish by ear very few differences stood out, but careful examination indicated trends that are compatible with my field classification into several types. In 44 recordings of seven vocalization types emitted by adults and the older among juveniles, the minimum sound frequencies are between 0.1 and 0.8 kHz, with an average of 0.5 kHz. Eighty-six percent of these values are in the range of 0.3 to 0.7 kHz. The maximum sound frequencies of the same vocalization are between 1.0 and 2.3 kHz, with an average of 1.5 kHz, and 86 percent of the values in the range of 1.1 to 2.0 kHz (Figure 11.5). Within the total frequency range of most vocalizations there is a narrower main frequency range imprinted more strongly on sonograms. Some sonograms cannot be said to contain a main frequency range as contrasted with the total range, as the vocalizations they represent appear devoid of weaker overtones, resulting in sonograms with approximately equal intensities throughout.

If the values of the differences between the minimum and the maximum sound frequencies of each of the various vocalizations are plotted against the maximum frequencies recorded for these vocalizations, a strong positive correlation ($r = 0.9$) is revealed (Figure 11.6). This is to

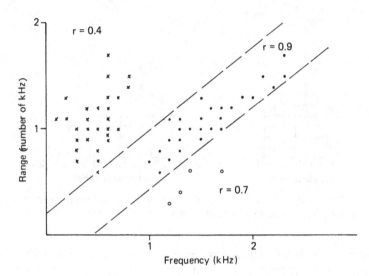

Figure 11.6. Individual vocalization. Sound frequency range of adult vocalizations compared to their maximum frequencies(●), with an indication of a trend. For comparison, minimum sound frequencies(X) of the same vocalizations and maximum frequencies for calves (O) are also plotted.

be expected if the minimum sound frequencies of all the vocalizations do not greatly differ from one another, whereas the maximum frequencies do so, which tends to be true of African buffalo vocalizations.

Vocalizations of the same type according to the field classification do not show identical maximum frequencies in sonograms. The variations in maximum frequencies within several vocalization types are plotted in Figure 11.7a (total range) and 11.7b (main range). A partial to complete separation of the types is indicated with respect to maximum frequency.

The minimum frequencies of calf vocalizations recorded by me are from 0.8 to 1.1 kHz, and the maximum frequencies from 1.2 to 2.3 kHz, but not exceeding 1.7 kHz in younger calves. These vocalizations are relatively few and show little variation, except that the range of frequencies increases with maturation. Calf vocalizations are plotted for comparison in Figures 11.6 and 11.7.

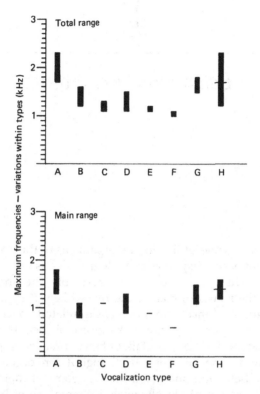

Figure 11.7. Individual vocalization. Variations in maximum sound frequencies within and between several buffalo vocalization types recognized in the field. Letters along the horizontal axis correspond to letters indicating vocalization types in the text. At H, the marker indicates the top of the variation in maximum sound frequencies for the younger calves. *Key:* (A) signal to move; (B) direction-giving signal; (C) water signal; (D) position signal; (E) warning signal; (F) aggressive signal; (G) mother-to-calf call; (H) calf distress call.

12

Behavioral background

Animal behavior is the utilization by organisms of their anatomical and neural capabilities in response to biological needs and environmental opportunities. The more two animals resemble one another, and the more similar their environments, the more likely they are to have behavior in common. Since comparatively unrelated species may largely resemble one another, behaviors, like morphological traits, cross taxonomic boundaries, and one can talk of behavioral convergence in much the same way as one talks of morphological convergence. Because at the same time behaviors may be inherited, some of their combinations are typical of groups of closely related genera. Thus logically, forms that are closely related, morphologically alike, and inhabit similar environments can be expected to show the greatest amount of similar behavior, though this is subject to modifications caused by general competition for food and habitat.

The bovid connection

Tendencies and influences held in common

How does *Syncerus caffer caffer* compare in behavior to other closely related species? For comparative purposes, some mention will be made in this chapter of a variety of bovids. This will include most members of Bovini, the tribe of large-boned cattlelike bovids in the genera *Bison*, *Bos*, *Bibos*, *Bubalus*, and of course *Syncerus*. It will include sheep and goats, members of the tribe Caprini, and the cattle-shaped arctic musk ox (Ovibovini). African alcepaphine antelopes – wildebeest and hartebeest – of the genera *Connochaetes* and *Alcelaphus*, respectively, will also be mentioned. These are large, characteristically high-withered and low-rumped, gregarious grazers, which inhabit plains and woodlands and, together with the much more distantly related tragelaphine eland, look most like bovines among the antelopes. Eland is the largest living antelope, whose resemblence to cattle is reflected by its generic name, *Taurotragus*. The other tragelaphines in this chapter – kudu (*Tragelaphus strepsiceros*) and situtunga (*T. spekei*) – are medium to large dwellers of closed habitats and members of a largely browsing genus. Those kobs (*Kobus*) that will be mentioned are medium-sized grassland and marshland antelopes that remain not far from water. Gazelles (*Gazella*) are fine-boned, light-bodied, small to medium, fast-running antelopes, which typically inhabit open dry plains. The impala (*Aepyceros melampus*) resembles the gazelle in size and conformation but tends toward more closed habitats. These bovids are normally horned, except for the females of sheep, goat, kudu, situtunga, kob (puku), and impala. All of the antelopes mentioned belong to African species, although the bovines, caprines, and ovibovines include also American, Asiatic, and European species.

Data available for different species vary in their completeness and in the kind of reporting. This hinders comparisons. Nevertheless, despite a still substantial lack of information regarding the behaviors of many species, enough is available to suggest that bovid behavior varies relatively little interspecifically. Most behavior observed to date in Bovini and Caprini consists of only a very few patterns, which are recurrent in different species, though variously combined. Many of these basic patterns are shared with species of other bovid tribes, for example, Ovibovini, Alcelaphini, Tragelaphini and other antelopes, as well as with the families Cervidae and Antilocapridae. Within most of these taxonomic groups body size varies substantially from species to species, imposing different survival requirements, and various member species favor quite different habitats, ranging all the way from high rocky mountains to marshes. Thus it is their comparatively close hereditary links, reflected by a number of morphological similarities, that may account for many similarities of behavior. They inherit several characters likely to influence behavior. Possibly most important among these, the ruminant alimentary system that these groups all possess must exert a powerful direct and indirect unifying influence on behavior, for two reasons: (1) All ruminants necessarily spend a very large part of their lives on similar alimentary processes and associated activity; and (2) the ruminant system normally induces the limitations of a strictly vegetarian life style. (One cannot leave out the word *normally* because, annoyingly, exceptions exist: the water chevrotain, *Hyemoschus*, a ruminant related to the Cervidae, and some African duikers (*Cephalophus*) partly feed on invertebrate and vertebrate animals and carrion. In the present context, however, these marginal exceptions can, I feel, be ignored.) Arboreal behavior is almost totally eliminated by the limb structure in all of these groups (though to be strictly accurate it is necessary to cite the not unskillful and habitual climbing of low-branched trees by the domestic Nubian goats, frequently observed by me in northern Sudan). None of these taxonomic groups contains any burrowing members. The very nearly general lack of arboreal and burrowing capability brings all these species closer to one another behaviorally, by this reduction of behavioral alternatives. The widespread occurrence of frontal transversely paired horns, though less general and/or uniform than the other attributes mentioned here, but at the same time restricted at present to ruminants, may also tend to generate behavioral similarities.

Sinclair (1977, p. 306) tabulated, for nine species of Bovini, 15 activities, all classified by him as elements of agonistic behavior, marked present (+), unknown (?), or with a minus sign (−), which must be assumed to indicate that the activities listed were expected (or confirmed) to be absent. The nine species (*S. c. caffer, Bubalus bubalis, Bison bison, B. bonasus, Bos grunniens, B. gaurus, B. banteng, B. sauveli, B. taurus*[1]) positively display from 7/15 (46.7 percent) to 11/15 (73.3 percent) of these activities, with a mean value of 9.2/15 (61.5 percent). The symbol "?" is used in 14 instances (i.e., 10.4 percent of all possible data). One modification to Sinclair's score for *S. c. caffer* is necessary since pushing fights, marked with a minus sign "−," have been observed by me in this species (Chapter 8). This brings *S. c. caffer* into the "top group," of 11/15 activities present (+), in company with *Bubalus bubalis* and *Bos (Bibos) gaurus*. In this type of compilation it is difficult to make a distinction between an "absent" (−) and an "unknown" (?) class, unless "present" (+) in one behavior necessitates "absent" (−) in another (which is not the case in this set of data). Therefore, the "present" (+) data have a different and more definite value than the other two categories (− and ?), which are somewhat unclear. The table indicates the following 5 of the 15 behaviors as common to all nine species of Bovini (using Sinclair's wording):

> Head up, present horns threat
> Lateral display
> Rubbing face in earth
> Earth tossing
> Horning bushes

My own observations of *S. c. caffer* include all the above behaviors, but although I have observed a form of "rubbing face in earth," I interpreted it as functional rather than ritualistic, apparently unlike Sinclair. These data well illustrate the recurrence of the same behavioral components in various bovid species. Several bovid behaviors extend into

---

[1] The last five species on the list are all placed by Sinclair (1977, p. 306) in the genus *Bos*.

other ungulate families, notably the Cervidae, for example ground-horning and horn displays, and flehmen (Chapter 9).

The greater part of all data that we possess on bovid behavior relates to domestic breeds, which thus provides the best standard for comparisons. *S. c. caffer* resembles domestic cattle of the genus *Bos* in a few behaviors, but in others seems closer to sheep. Indeed, the behavior of domestic and feral sheep (*O. aries*) has some resemblences to that of both genera of living buffaloes, *Bubalus* and *Syncerus*. Differences in behavior can be explained in some cases by the dissimilarity of habitats, but in other cases this explanation seems insufficient. This is exemplified further on in this chapter, in a comparison of the African buffalo of the Busanga area, Zambia with the water buffalo (*B. bubalis*) of northern Australia.

Grazing

Comparisons of everyday behaviors are sometimes complicated by the multiplicity of modifying factors as well as differences in the reporting of equivalent data for various species. Grazing belongs in this class. It is adjusted to prevailing seasonal and regional conditions, which greatly increases the number of variables to be considered when making comparisons. Grazing and ruminating are interdependent. Ruminating is affected by the water content and texture of the grass intake. These are highly variable and thus differences in the length and distribution of grazing and ruminating bouts should be expected, not only between different species, which may be equipped with jaws of different sizes and may require different amounts of food, but within single species as grazing conditions change from one season or locality to another. The time distribution of individual grazing bouts does in fact differ on a daily basis. However, some regularities can be detected. All observed bovid species tend to graze intensively for at least 1 hour (1) either at dawn or in the early morning, and (2) either in the late afternoon or around sunset. The adjective *intensively* is used with caution in the case of alcelaphine antelopes, which tend to graze at a uniform rate. Finally, (3) the time range spent grazing during each 24-hour period can be determined for different species. Table 12.1 is an effort to present a few grazing data in comparable terms for five bovid species.

Mating

Bovid mating behavior can be divided into three categories, one with a territorial component and two without (although of course a home range may exist).

Table 12.1. *Some basic grazing data for five bovid species*

Species (source)	Dawn	Early morning	Mid-morning	Mid-day	Mid-afternoon	Late afternoon	Sunset	Night	Total daily time (hr)
B. taurus (Hafez, 1975; own observation)	X̲	x	X̲	x	X̲	x	x	0/*	4–9
O. aries (Hafez, 1975; own obs.)	0	X̲	x	x	x	x	x	0	9–11
B. bubalis (Tulloch, 1978)	X̲*	X̲	X̲	0	X̲	X̲	X̲	0/**	6–12
S. c. caffer (own obs.)	x	X̲	x	0	0/L	X̲	x	X̲	6–11
A. lichtensteini (Dowsett, 1966; own obs.)	x	X̲	x	L	L	X̲	x	X̲	?

Key: X̲ = typical grazing period; x = grazing occurs but no definite pattern; L = typically little grazing; 0 = typically no grazing; * = grazing only in dry season/ hot sunshine months; ** = grazing during abnormally long drought only.

(1) The male acquires a territory, marks it, holds it against other adult males, and mates with females that enter it, which he may or may not retain within the territory. This pattern occurs in blue wildebeest (*Connochaetes taurinus:* Estes, 1969), Lichtenstein's hartebeest (*Alcelaphus lichtensteini*), puku (*Kobus vardoni*), and others.

(2) A male acquires high status as compared to other males, collects a "harem" of females, numbering from 1 to perhaps 100, and defends it against other adult males, as in impala (*Aepyceros melampus*), lesser kudu (*Tragelaphus imberbis*), and so on.

(3) A male, which may be one of several tolerated within a group, attaches himself to one proestrous or estrous female at a time (she is normally one of a group) and eventually mates with her, unless displaced by a higher-status male. This is a common bovine pattern, and one that occurs in the African buffalo. All three categories are amenable to further subdivisions.

Bovid morphology compels males to mount the females from the rear, the anterior parts of a male's body over a female's hind quarters (Chapter 9). Any minor interspecific variations in this basic copulatory posture can be explained by the size and weight differences of the participants.

Bovid behavioral sequences related to mating range interspecifically from moderately complexly ritualized to fairly basic. Some component behaviors are widespread, as for example flehmen, which is encountered in a variety of ungulates. Although I question whether flehmen is restricted to the mating ritual in the African buffalo (Chapter 9), this is certainly one of its applications. The raised-front-leg posture of the male, some part of the leg touching the female's body and named the *"laufschlag,"* is recurrent in smaller African bovid species, including at least some gazelles and kobs. This behavior has not been observed in *S. c. caffer* but is mentioned here because it is occasionally combined with another, simultaneous posture of the male, which consists of resting the bottom of the lower jaw on the female's back, while "supporting" her ventral region with a front leg in the "laufschlag" posture. This has been observed in the Uganda kob (*K. k. thomasi*) pre- and intercoitally, and reported as also a postcoital behavior by Buechner and Schloeth (1965). Walther (1964) reports this behavior in the Dorcas gazelle (*G. dorcas*). One may speculate that this behavior is related to the precoital chin-resting of the bovines, including *S. c. caffer* (Chapter 9). Male precoital behavior very similar to the bovine chin-resting has also been seen in eland (*T. oryx*) in Zimbabwe and in situtunga (*Tragelaphus spikei*) in Zambia.

Group defense

A defensive formation is used by musk oxen (*Ovibos moschatus*)against wolves (e.g., Tener, 1954). It consists of a closely packed array of fully

Figure 12.1. A group of bachelor males on the defensive as lions approach from behind me and my right. The middle animal takes a step forward, to investigate, as he spots me between him and the lions. (Narok District, Kenya.)

grown animals, which forms a protective barrier between the predators and the calves. A similar defensive formation has also been reported in feral water buffaloes (*Bubalus bubalis*) from Sri Lanka. (Eisenberg and Lockhart, 1972) and can be occasionally seen in African buffalo groupings. In threatened herds of *S. c. caffer* a less pronounced version of this behavior can be seen, when a few females move forward on the herd's endangered side and most males advance in front of the females in a close, though not shoulder-to-shoulder, formation. I have observed more extreme displays of similar behavior in threatened bachelor clubs of *S. c. caffer*, in Zambia and Kenya (Chapter 7; Figure 12.1). Adult musk oxen and buffalo males occasionally advance, singly or with one or two others, out of the defensive line to threaten or attack the predators.

A *S. caffer* and a *B. bubalis* population compared

Of all the bovids now living, the Asiatic water buffalo (*B. bubalis*) is morphologically closest to *S. c. caffer*, and the two are not far removed taxonomically. In the first half of the last century, water buffaloes were brought from Asia to northern Australia, and now constitute a large feral population, which has been worked and reported on by Tulloch (1975, 1978, 1979). The habitat of Tulloch's water buffaloes and that of the *S. c. caffer* population that I observed in the Busanga flats of western Zambia happen to be much alike, and thus a comparison of the behaviors of these two populations is of interest.

Both areas are largely floodplains, intercalated with low wooded ridges. The floodplains are inundated during an annual wet season, which in the Australian region lasts from September/October to March/ April, and in the African region from October/November to April/May. Both have a rainless dry season when water becomes gradually restricted to deep channels, and in both regions a perennial grassland flora is burnt during the dry season. Both are similarly situated south of the equator, one at about 12½°S and the other at 14½°S. The greatest discrepency may be in the temperatures, and even that difference is moderate: in the Australian region temperatures range from 20 to 40°C (Tulloch, 1978) and in the African region from 9 to 31°C (from averages measured at Lusaka). The hottest and coolest periods of the year also coincide.

Whereas the physiographic and climatic factors are largely similar in the two regions, there is no predation upon the Australian water buffaloes analogous to lion predation on the African buffaloes of the Busanga. The movements of both populations are relatively unhindered.

The Australian water buffaloes are not active intermittently throughout the 24-hour cycle but, like sheep, camp overnight on the same site, to which they walk at sunset and which they leave at dawn to graze, along the same well-worn paths, whereas the Busanga buffaloes alternately graze and rest both day and night and normally occupy a different resting site each time. Whereas the water buffaloes go to a "dung heap" to defecate, the Busanga and other African buffaloes do it at their resting site (Chapter 6). The water buffaloes wallow frequently and in groups, with social facilitation possibly an important factor, whereas the Busanga buffaloes wallow only occasionally and singly or in small incidental parties.

The social divisions in Tulloch's water buffalo population appear to have some parallels with the social division in *S. c. caffer* herds, including those of the Busanga. Tulloch (1978) lists for the water buffaloes:

Clan, that is, family party, which may consist only of a female and her offspring, or may include other generations, totaling as many as 30 individuals and comprising subadult males of up to 3 years of age.
Group, that is, a number of clans that all camp at the same spot, know one another, and share one home range.
Herd, that is, a number of adjacent groups, whose members do not know one another well, but whose home ranges touch or overlap, and which may share the same facilities, for example water, pasture, or woodland.

Adult *B. bubalis* males live separately, in home ranges of their own, and associate with conspecifics only during the mating season. *Groups* belonging to the same *herd* occasionally meet and after an elaborate ceremonial stay together amicably for part of the day. (Italics indicate Tulloch's sense of these terms.) Home ranges are "stable." Tulloch reports that in 4 years on one occasion only – during an exceptionally long dry season that caused high mortality–were the home ranges abandoned. This drought, however, was a catastrophe that led to partial social collapse and subsequent restructuring, and is only of subordinate interest in the present context, which deals with more ordinary situations.

The *S. c. caffer* basic herd is a broader unit than the *clan*, as it appears to contain individuals that are unrelated or only remotely related. I consider it the smallest fully viable social unit of *S. c. caffer*, as it contains or is accompanied by some males capable of mating, although its basis, as in the water buffalo *clan*, is a female hierarchy (Chapter 4). The *clan* concept could be applied within basic herds of *S. c. caffer*, but these do not normally split into lesser functional social units. The smallest groupings in some populations of *S. c. nanus* probably resemble the *clan*, with the difference that adult males may be associated in them at any time (i.e., they also correspond to my definition of basic herd).

Tulloch's water buffalo *group* is very near to my concept of herd in *S. c. caffer*.[2] The difference appears to be largely that of size, that is, a herd of *S. c. caffer* often comprises 300 to 500 individuals, whereas Tulloch's (1978) data suggest that the water buffalo *group* may not attain 300. The average of five annual counts of four different *groups* (Tulloch, 1978) is 129.3, varying from 34 to 261. But apart from this, and with the difference that some adult males are normally present in *S. c. caffer* herds, the structure and degree of social cohesion of a water buffalo *group* appear equivalent to those of a *S. c. caffer* herd.

A water buffalo *herd* resembles in most respects a composite herd of *S. c. caffer*, that is, it is an entity that exists much of the time in a fragmented state, but if it comes together, on occasions when environmental conditions permit, a moderate level of social integration is present.

Although, by contrast to the Australian water buffalo, a few adult males are always present in social groupings of *S. c. caffer*, separate bachelor clubs also exist (Chapters 4 and 7).

Perhaps we can say about the social behaviors of the Australian water buffalo and *S. c. caffer* that we are looking at the same pattern, modified only to suit (1) a sedentary lifestyle in the former and a nomadic one in the latter species; and (2) the presence of a sharply defined mating season in the former, as contrased with year-round mating (though often with a prominent peak) in the latter.

---

[2] By coincidence, I originally proposed the same term, *group* (e.g., unpublished notes to the Zambian National Parks & Wildlife Service), reserving the term *herd* for the largest observed social unit of *S. c. caffer* that has relatively little permanence and a fairly low level of mutual recognition among individuals. I later altered the nomenclature, wishing, for reason of systematics, to retain the word *herd* in all types of viable social units of *S. c. caffer*, and proposed the nomenclature that I now employ, i.e., basic herd, herd, and composite herd.

Buffalo (*S. c. caffer*) and eland (*T. oryx*) behaviors compared

At this stage, and hot on the heels of the comparison made between the two living large buffaloes (*S. c. caffer* and *B. bubalis*), which are rather closely related forms placed in one bovid tribe, it is of interest to look at an example of behavioral convergence between members of two quite distinct bovid tribes – the African buffalo of the *caffer* subspecies and the eland (*T. oryx*). The eland belongs to the Tragelaphini, a tribe of browsers, sedentary bush/woodland or reed swamp dwellers, which live singly or in small family groups, and whose behavior is well adapted to this mode of existence. The eland, however, has adopted a grasslands lifestyle. It has a taurine body conformation, lives in herds that may number hundreds of individuals, and has developed the nomadic behavior necessary in large gregarious grazers, which require large volumes of food in sometimes water-defficient and poor habitats. It has, however, retained a tragelaphine liking for forbs, shrubs, and tree foliage in its diet (Hofman and Stewart, 1972), and hence a tolerance for a wide range of habitats besides grasslands. Like *S. c. caffer*, it is not a species-selective grazer. Several high-status males are present simultaneously within eland herds, and separate bachelor clubs occur. Antipredator defensive arrays, similar to those already described for the African buffalo and other bovids, have been observed, though rarely, and there are no reports of territorial behavior in eland herds (Kruuk, 1972; Jarman, 1974). In these respects it resembles *S. c. caffer*. The herds are much wider-ranging than those of buffaloes, since their movements are not limited by the need to drink every day, and thus need not adhere to a home range in any but the broadest sense. Eland appear less aggressive than buffaloes, despite their large size, weight, and functional horns present in both sexes, possibly because their greater speed makes flight the more efficient choice in almost all cases of interspecific confrontation (also see the section in Chapter 8, "Antipredator behavior"). Eland and buffalo are encountered, and are successful, in the same African regions. The eland's tolerance for greater water-to-pasture distances, its greater preference for browse, and its greater mobility effectively reduce competition between the two forms.

Home range and family group territory

Bovid home ranges can be seasonal, that is, routinely abandoned and reoccupied at particular times of the annual cycle. The present-day classical example is undoubtedly the seasonal wildebeest migration in the Serengeti–Lake Victoria region of northern Tanzania. When home ranges are not vacated seasonally but retained until some crisis compels or promotes abandonment they are often called "stable" or "permanent," for example the home ranges of the northern Australian water buffaloes. However, spacing between crises and their intensity may vary widely , and stability is seldom – and permanence is never – a ruling attribute of these home ranges. I therefore prefer to call this type of home range "nonseasonal" or "long-term."

Despite their nomadic lifestyle, *S. c. caffer* herds remain within home ranges. These, though not often abandoned and reoccupied seasonally, may be only partly used in certain seasons, and the extent to which various parts are used may undergo seasonal changes. The home range boundaries may be observed to shift more or less markedly over a period of a few years, and the area of greatest occupancy within a home range also may shift. The home ranges of African buffalo herds are nomally larger than home ranges occupied by water buffalo *groups*, as described by Tulloch. Using his (1978) mean densities and *group* counts, the water buffalo home ranges appear to extend over areas of the order of 10 square kilometers, perhaps less. Domestic sheep, even when free-roaming, remain within home ranges that in the Scottish hills attain 120 acres (Hunter and Davis, 1963) or 5 square kilometers. Home ranges of *S. c. caffer* herds vary from some 100 to some 1000 square kilometers. Despite size differences, the home ranges of both buffalo genera and of sheep are similar in their propensity to overlap, that is, boundaries are indefinite and unguarded, and neighboring social units are tolerated within each other's home range.

This type of home range is in sharp contrast with the family territories of hartebeests, for example, *Alcelaphus lichtensteini* (Dowsett, 1966) or (sometimes) the blue wildebeest, *C. taurinus* (Estes, 1969). The family group territories of Lichtenstein's hartebeests are marked, defended, and do not overlap. In large populations of blue wildebeest, a very large social entity exists, or periodically forms, that has parallels with the *S. c. caffer* composite herd but may number many thousands, and is consequently more fragmented. It occupies either a long-term or seasonal home range.

The American bison's ecological niche, before the drastic reduction of its numbers in the 1870s and 1880s, appears to have resembled that of the wildebeests in Africa, more than the probably closer related African buffaloes. The enormous bison herds migrated annually between a northern and a southern home range in such great numbers that, like wildebeests, they had a direct and marked modifying effect on the whole environment (Roe, 1970) by generating subsidiary movements of other species (e.g., predators), altering vegetation growth by mass trampling, manuring, and reseeding, and even altering minor topographic and drainage features. The bison home ranges were vast and defined only by climato-physiographic boundaries.

Bachelor clubs

The separate or semiseparate existence of male bachelor clubs is a widespread bovid pattern. Bachelor clubs, very variable as to the number of members, may be nonterritorial and free-roaming (e.g., *C. taurinus*) or may have mobile territories (e.g., *A. lichtensteini*). There is competition between members of all bovid bachelor clubs, but the meaning of the competition varies. Bachelors of the two alcelaphine species mentioned here compete for future roles that exclude all but one male at a time, that is, family group leadership in hartebeests, or mating territory occupation in the blue wildebeest, whereas competition within the bachelor clubs of Bovini/Caprini is for ranking in future behavioral interactions in which several males are tolerated simultaneously. Bachelor clubs of American bison, a species with a well-defined mating season, behave similarly to those of *B. bubalis*, returning to the herds to rut, although as in *S. c. caffer* a few adult males may stay with the herds permanently or semipermanently.

Shifts of emphasis in the meaning of similar behaviors

The shift in the meaning of outwardly similar movements or postures from one species to another is one of the more interesting aspects of behavior within closely related species and genera. The same behavior that in some species appears to serve mainly for grooming functions in others as a direct threat or a territorial display. There is a shift at least in the emphasis, if not the full meaning, of the activity, from one to another species. In this section, several behaviors are described with a view to illustrating the above observation.

Within the blue wildebeest's large home range, undefined by sharp boundaries, individual males establish and defend small well-defined territories of less than 20 to some 150 meters in diameter. Female-centered nursery herds pass through these territories in the course of their ordinary daily movements, and any estrous female may then be covered by the resident male (Estes, 1969). This behavior appears to be absent in the American bison. An interesting possible link between the bison and the wildebeest is the dry wallow of the former (McHugh, 1958; Roe, 1970) and the stamping ground of the latter (Estes, 1969). Both are patches of bare ground on which males urinate, defecate, roll, and wallow – behaviors that appear to contain elements of agonistic displays. In the blue wildebeest, the stamping ground is situated within a male's mating territory; in the American bison, the dry wallow is less restricted and does not appear to be part of any territorial behavior. A behavioral link also appears possible between the wildebeest stamping ground and the territorial demarcation patches of Lichten-

Table 12.2. *Main significance of some similar and possibly related behaviors in six bovid species*

Behavior	Species	Main significance
Dry wallow	*Bison bison*	?G/C
Stamping ground	*C. taurinus*	C/T
Marking patches/kneeling/ ground-horning	*A. lichtensteini*	T/?C
Kneeling/ground-horning	*Bos taurus*	C
Kneeling/ground-horning	*C. taurinus*	C
Mud wallow/mud-horning	*S. c. caffer*	G/?C
Mud wallow	*Bubalus bubalis*	G

Key: C = nonterritorial threat or challenge; T = territorial behavior; G = grooming behavior.

stein's hartebeest. There are several of these in prominent places within a hartebeest family group territory. They are bare patches of ground, measuring about 1.5 to 4.5 feet square (Dowsett, 1966) or some 0.2 to 2 square meters, made by hartebeests of both sexes, from a kneeling position, by horning the ground and then rubbing the cheeks over it, which may be a way of depositing some of the abundant preorbital gland secretion present in *A. lichtensteini.* Now it is of interest to recall that horning the ground in a kneeling position is a classical challenge display of the domestic bull, *B. taurus* (e.g., Fraser, 1974) and of *C. taurinus.* The African buffalo's wallowing behavior does not appear to include a fully ritualized agonistic display (Chapter 8), although it probably does sometimes contain a component of such a display, possibly stimulated by the similarity of movements involved in mud-covering the head (for functional reasons) and aggressive horning. An example of this kind of stimulus is found in the escalation of head-rubbing against trees to aggressive head pushing and horning (Chapter 8). When a *S. c. caffer* bull horns the muddy soil of a wallow, he often does it in a kneeling position reminiscent of the challenge display in *Bos taurus* males. The position, however, is achieved from a recumbent (wallowing) posture, and not from a standing one as in the *B. taurus* display. This distinction may be of some importance, as it suggests that in the domestic bull the kneeling posture is achieved by special design, whereas in the buffalo bull it may be merely incidental to getting up. In Table 12.2 this series of possibly related behaviors is condensed.

The African connection

Are any of *Syncerus caffer*'s behaviors likely to be specifically African adaptations? Possibly its relationship to water belongs in this category.

If we accept that there is a high degree of similarity between the Asiatic water buffalo and the African buffalo, it seems necessary to explain the comparative scarcity of group water-entering and wallowing behavior in the African genus, whereas it is so habitual in the Asiatic form. The African buffalo, like the Asiatic water buffalo, shows a preference for river valleys, lake basins, and floodplains, which makes the discrepency in water-entering and wallowing behavior seem even more anomalous. As hinted in Chapters 6 and 8, the reason may be crocodiles. Until recently, most rivers, lakes and deeper swamps of sub-Saharan Africa had large Nilotic crocodile populations. It seems almost certain that a regular habit of entering water and wallowing in all members of a plentiful bovid species, such as was *S. caffer*, would make it a regular crocodile prey. Being regularly available in abundance to the crocodile, the buffalo easily could have become a preferred prey and started attracting crocodiles. This is why the group water-entering and wallowing behavior may have failed to develop or become lost. Since the only opportunity for safe wallowing was in small inland water pans which could not accommodate more than a few individuals, wallowing may have become and remains a sporadic behavior. Although the high-status members of an African buffalo herd are not the only individuals that wallow, they appear to do so much more often than less-dominant individuals. It seems reasonable that if the supply of wallows is less than the demand for them, occupancy of a wallow could become a status sign. Asiatic crocodiles of species feeding on large land mammals were less prolific and less widely distributed, at least in the less remote times. The African buffalo's dislike of making long rest halts on soggy ground is not incompatible with the hypothesis just advanced.

One may extend the argument and suggest in general that the comparatively smaller dependence on water is an African adaptation, due to high pressures on waterside pastures from hippopotamus and numerous other ungulates and abundance of good graze at some distance from water. The fact is that, despite its liking for floodplains, *S. c. caffer* is a very successful African bush/grassland form, capable of doing well on one drink a day.

In conclusion, I hope that this traverse across the field of bovid behavior has suggested the order of similarity between the behavior of *Syncerus caffer caffer* and its cofamilials and left a hint of the kinds of behavioral limitations and alternatives available to the bovid.

References

The following abbreviations are used to denote references pertaining to specific areas of studies: cli. = climatology; pal. = paleontology (*Syncerus* lineage); r.t. = related topic (*S. caffer* usually not mentioned).

Agnew, A. D. Q. 1968. Observation on the changing vegetation of Tsavo National Park (East). *E. Afr. Wildl. J.*, 6. (r.t.)

Akeley, C. E. 1921. Hunting the African buffalo. *World's Work*, 41: 497–504.

Alexander, R. McN., V. A. Langman, and A. S. Jayes. 1977. Fast locomotion of some African ungulates. *J. Zool.* (London), 183:291–300.

Allen, G. M. 1939. A checklist of African mammals. *Bull. Mus. Comp. Zool.* Vol. 83. Cambridge, Mass.

Anderson, G. D., and L. M. Talbot. 1965. Soil factors affecting the distribution of the grassland types and their utilisation by wild animals on the Serengeti Plains. *J. Ecol.*, 53. (r.t.)

Ansell, W. F. H. 1960a. *Mammals of Northern Rhodesia.* Lusaka: Government Printer.

– 1960b. The breeding of some larger mammals in Northern Rhodesia. *Proc. Zool. Soc.* (London), 134(2):251–74.

– 1969. A multiple kill by lions and a stolen kill. *Puku*, no. 5:214–15.

– 1971. Order Artiodactyla. In *The Mammals of Africa: An Identification Manual*, J. Meester and H. W. Setzer (eds.), part 15. Washington, D.C.: Smithsonian.

Antonius, O. 1935. Zur Weiterentwicklung and Systematik der Afrikanischen Büffel. *Der. Zool. Garten J.*, 58, Heft 10–12:265–70.

Arnold, A. J. 1899. In *Great and Small Game of Africa*, H. A. Bryden (ed.). London: Rowland Ward.

Astley Maberly, C. T. 1960. *Animals of East Africa.* Cape Town: Howard Timmins.

– 1963. *The Game Animals of Southern Africa.* Cape Town: Howard Timmins.

Babault, G. 1949. Notes ethologiques sur quelques mammifères africains (suite). *Mammalia*, 13:65–66.

Baudenon, P. 1952. Notes sur les bovidés du Togo. *Mammalia*, 16:49–61.

Bainbridge, W. R. 1969. A parti-coloured buffalo. *Puku*, no. 5:216. (With photograph.)

Basilio, A. 1962. *La vida animal en la Guinea Espanola*, 2nd ed., pp. 132–8. Madrid.

Basson, P. A., et al. 1970. Parasitic and other diseases of the African buffalo in the Kruger National Park. *Onderspoort J. Vet. Res.*, 37:11–28.

Bate, D. M. A. 1949. A new African fossil long-horned buffalo. *Ann. Mag. Nat. Hist.* (London), 12(2):396–8.(pal.)
- 1951. The mammals from Singa and Abu Hugar. *Brit. Mus. Nat. Hist., Fossil Mammals of Africa*, no. 2:1–28. (pal.)
Beland, O. P. 1952. How long do they live? *Nat. Hist.* (New York), 61:131–4, 141–2.
Bell, R. H. V. 1970. The use of the herb layer by grazing ungulates in the Serengeti. In *Animal Populations in Relation to Their Food Resources*, A. Watson (ed.). Oxford: Blackwell. (r.t.)
Bell, W. D. M. 1960. *Bell of Africa*, pp. 31–2, 149–54, 184, 210, 212, 222–3. London: Holland Press.
Belon, P. 1555. *Les observations de plusieurs singularités et choses memorables trouvées en Grèce, Asie, Arabie et Egypte*. Enlarged ed. Antwerp. (Orig. publ. 1553).
Benson, C. W. 1969. Large mammals of the Liuwa Plain and Sioma-Ngwesi Game Reserves, Barotse. *Puku*, no. 5:49–57.
Berry, P. S. M. 1969. Wanton killing by young lions. *Puku*, no. 5:218–19.
Bertin, L. 1949. *La vie des animaux*, vol. 2, pp. 348–9.
Bigalke, R. C. 1972. The contemporary mammal fauna of Africa. In *Evolution, Mammals, and Southern Continents*, A. Keast, F. C. Enk, and B. Glass (eds.), part 4, pp. 141–94. Albany, N.Y.: State University of New York Press.
- 1978a. In *Biogeography and Ecology of Southern Africa*, M. J. A. Werger (ed.), vol. 2, chap. 31, pp.1026–7. The Hague: Dr. W. Junk bv Publishers.
- 1978b. *Taurotragus oryx*. In *Biogeography and Ecology of Southern Africa*, M. J. A. Werger (ed.), vol. 2, chap. 31, pp. 1030. The Hague: Dr. W. Junk bv Publishers. (r.t.)
Bigourdan, J., and R. Prunier. 1937. *Les mammifères sauvages de l'ouest africain et leur milieu.* Paris.
Bindernagel, J. A. 1971. *Elaeophora poeli* (Nematoda: Filaroidea) in African buffalo in Uganda, East Africa. *J. Wildl. Dis.*, 7:296–8.
- 1972a. Liver fluke *Fasciola gigantica* in African buffalo and antelopes in Uganda, East Africa. *J. Wildl. Dis.*, 8:315–17.
- 1972b. *Thelezia rhodesi* (Nematoda: Spiraroidea) in African buffalo in Uganda, East Africa. *J. Parasitol.*, 58:594.
Bindernagel, J. A., and A. C. Todd. 1972. The population dynamics of *Ashworthius lerouxi* (Nematoda: Trichostrongylideae) in African buffalo in Uganda. *Br. Vet. J.*, 128:452–5.
Blancou, L. 1935. Buffles de l'Oubangui-Chari-Tchad. *Terre Vie*, 2me sem., 6:202–7.
- 1948. Notes sur les mammifères de l'Oubangui-Chari. *Mammalia*, 1–2:2–3.
- 1954. Buffles d'Afrique. *Zooleo.*, n.s. 27:425–34.
- 1958a. Distribution geographique des ongules d'Afrique Equatoriale Francaise en relation avec leur ecologie. *Mammalia*, 22:294–316.
- 1958b. Note sur le statut actuel des ongules en Afrique Equatoriale Française. *Mammalia*, 22:399–405.
Bligh, J., and A. M. Harthoorn. 1965. Continuous radiotelemetric records of the deep body temperature of some unrestrained African mammals under near-natural conditions. *J. Physiol.*, 176:145–62.
Blyth, E. 1866. A note on African buffaloes. *Proc. Zool. Soc.* (London) 371–3. (with woodcuts.)
Bolton, M. 1973. Notes on the current status and distribution of some large mammals in Ethiopia (excluding Eritrea). *Mammalia*, 37:562–86.
Bouet, G. 1934. *Contribution a l'étude de la repartition des grands mammifères en Afrique française (Bovideés-Tragulidés)*, pp. 11–12. Paris: Societé d'Editions Geographiques, Maritimes, et Coloniales.
Bourgoin, P. 1955. *Animaux de chasse d'Afrique*, pp. 82–6. Paris: Toison d'Or.

Bourlière, F. 1963a. Observations on the ecology of some large African mammals. In *African Ecology and Human Evolution*, F. C. Howell and F. Bourlière (eds.), pp. 43–54. Chicago; Aldine.

– 1963b. Specific feeding habits of African carnivores. *Afr. wildl.*, 17(1):21–7. (r.t.)

– 1965. Densities and biomasses of some ungulate populations in eastern Congo and Rwanda with notes on population structure and lion/ungulate ratios. *Zool. Afr.*, 1.

Bourlière, F., E. Minner, and R. Vuattoux. 1974. Les grands mammiferes de la region de Lamto, Cote d'Ivoire. *Mammalia*, 38:433–47.

Bourlière, F., and J. Verschuren. 1960. *Introduction a l'ecologie des ongules du Parc National Albert*. Brussels: Institut des Parcs Nationaux du Congo Belge,

Branagan, D., and J. A. Hammond. 1965. Rinderpest in Tanganyika: A review. *Bull. Epizootic Dis. Afr.*, 13:225–46.

Brand, D. J. 1963. Records of mammals bred in the National Zoological Gardens of South Africa during the period 1908–1960. *Proc. Zool. Soc.* (London), 140(4):617–59.

Brentjes, B. 1969. *African Rock Art*. London: Dent. (pal.)

Brocklehurst, H. C. 1931. *Game Animals of the Sudan*. London: Gurney & Jackson.

Brocklesby, D. W. 1965. A new theilarial parasite of the African buffalo (*Syncerus caffer*). *Bull. Epizootic Dis. Afr.*, 13:325–30.

Brocklesby, D. W., and S. F. Barnett. 1966. The isolation of *Theileria lawrencei* (Kenya) from a wild buffalo (*Syncerus caffer*) and its serial passage through captive buffaloes. *Brit. Vet. J.*, 122:387–95.

Brooke, Sir V. 1873. On African buffaloes. *Proc. Zool. Soc.* (London), part 2:474–84.

– 1875. Supplementary notes on African buffaloes. *Proc. Zool. Soc.* (London), part 3: 454–7.

– 1875. *B. pumilus*. *Ann. Nat. Hist.* (4) 13:156–60.

Browne, W. G. 1799. *Travels in Africa, Egypt, and Syria*. London: Cadell, Davies, Longman, and Rees.

Bryden, H. A. 1899. In *Great and Small Game of Africa*, H. A. Bryden (ed.). London: Rowland Ward. Entries by Selous, Jackson, Penrice, Arnold.

Brynard, A. M. 1968. The influence of veld burning on the vegetation and game of the Kruger National Park. *Ecol. Stud. S. Afr.* (r.t.)

Buckley, T. E. 1877. On the past and present geographical distribution of the larger mammals of South Africa. *Proc. Zool. Soc.* (London), 452–6.

Buechner, H. K., and R. Schloeth. 1965. *Z. Säuget.* (r.t.)

Burton, M. 1962. *Dictionary of the World's Mammals*, pp. 208–9. Great Britain: Museum Press.

Butzer, K. W. 1963. Climatic-geomorphologic interpretation of Pleistocene sediments in the Eurafrican subtropics. In *African Ecology and Human Evolution*, F. C. Howell and F. Bourlière (eds.), pp. 1–27. Chicago: Aldine. (cli.)

Cansdale, G. S. 1946. *Animals of West Africa*, p. 26. London: Longmans, Green & Co.

– 1948. *Provisional Check List of Gold Coast Mammals*, p. 14. Accra: Government Printer.

Carbou ("In Tanoust"). 1930. *La chasse dans les pays saharien et sahelien de l'A.O.F. et de L'A.E.F.* Paris.

Child, G., P. Graham, and C. R. Savory. 1964. The distribution of large mammal species in Southern Rhodesia. *Arnoldia* 1(14).

Child, G., P. Smith, and W. von Richter. 1970. Tsetse control hunting as a measure of large mammal population trends in the Okavango delta, Botswana. *Mammalia*, 34:34–75.

Christy, C. 1929. The African buffaloes. *Proc. Zool. Soc.* (London), 445–62.

Churcher, C. S. 1972. Late Pleistocene vertebrates from archaeological sites in the Plain of Kom Ombo, Upper Egypt. *Contrib. Life Sci. Div. Roy. Ont. Mus.*, 82:1–172.(pal.)

– 1974. Relationships of the Late Pleistocene vertebrate fauna from Kom Ombo, Upper

Egypt. In *Contributions to the Paleontology of Africa*, Rushdi Said and B. H. Slaughter (eds.). *Proc. 75th Anniv. Geol. Surv. of Egypt*, 4:363–84. (pal.)

– 1980. Dakhleh Oasis Project – Preliminary observations on the geology and vertebrate palaeontology of northwestern Dakhleh Oasis: a report on the 1979 fieldwork. *SSEA J.*, 10(3):379–95.(pal.)

– 1981. Dakhleh Oasis Project – Geology and palaeontology:interim report on the 1980 field season. *SSEA J.*, 11(4): 191–211. (pal.)

Cincinn, J. 1887. Description of *B. aequinoctialis*. Bull. Mens. Soc. Nat. d'Acclim (Paris), p.86.

Clark, J. D. 1960. Human ecology during Pleistocene and later times in Africa south of the Sahara. *Curr. Anthrop.*, 1(4):307–24. (cli.)

Cloudsley-Thompson, J. L. 1967. *Animal Twilight*. London: Foulis.

Cooke, H. B. S. 1949. Fossil mammals of the Vaal River deposits. *Mem. Geol. Surv. S. Afr.*, 35(3):1–117.(pal.)

– 1962. The Pleistocene environment in Southern Africa: hypothetical vegetation in Southern Africa during the Pleistocene. *Ann. Cape Prov. Mus.*, 2:11–15. (cli.)

– 1972. The fossil mammal fauna of Africa. In *Evolution, Mammals, and southern Continents*, A. Keast, F. C. Erk, and B. Glass (eds.), Albany, N.Y.: State University of New York Press. part 3, pp. 89–140. (pal.)

Cooke, H. B. S., and S. C. Coryndon. 1970. Pleistocene mammals from the Kaiso Formation and other related deposits in Uganda. In *Fossil vertebrates of Africa*, L. S. B. Leakey and R. J. G. Savage (eds.), vol. 2, pp. 107–224. London: Academic Press. (pal.)

Cornwallis Harris, W. 1840 *Portraits of the Game and Wild Animals of Southern Africa*. London. (Reprint: Cape Town: Balkema, 1969, pp. 81–87).

Crandall, L. S. 1964. *The Management of Wild Mammals in Captivity*. Chicago: University of Chicago Press. (r.t.)

Cross, M. W., and V. J. Maglio. 1975. *A Bibliography of the Fossil Mammals of Africa, 1950–1972*. Princeton, N.J.: privately printed. (pal.)

Cullen, A. 1969. *Window onto Wilderness*. Nairobi: East African Publishing House.

Curry-Lindhal, K. 1961. *Exploration du Parc National Albert et du Parc National de la Kagera*, fasc. 1, 123–7. Brussels: Institut des Parcs Nationaux du Congo et Ruanda-Urundi.

Dagg, A. I., and A. Taub. 1970. Flehmen. *Mammalia*, 34:686–95.(r.t.)

Dalimier, P. 1955. *Les buffles du Congo Belge*. Brussels: Institut des Parcs Nationalaux du Congo Belge.

Dalquest, W. W. 1965. Mammals from the Save River, Mozambique, with descriptions of two new bats. *J. Mamm.*, 46:254–64.

Dane, R. M. 1931. Methods of attack by buffalo. *Field*, 158-4102:219.

Darling, F. F. 1960. *An Ecological Reconnaissance of the Mara Plains in Kenya Colony*. Wildlife Monographs 5. Nairobi. (r.t.)

Dasmann, R. F., and A. S. Mossman. 1962. Reproduction in some ungulates in Southern Rhodesia. *J. Mamm.*, 43:533–7.

Day, D. 1969. Killing of a lioness by buffalo. *Puku*, no.5:219.

Decken, Baron C. C. von der. 1869. *Reisen in Ost-Africa*, Leipzig.

Dekeyser, P. L., and J. Derivot. 1957. *Etude d'une collection de têtes osseuses de buffles de l'Afrique Noire*. Dakar.

Dekeyser, P. L., and A. Villiers. 1951. *Animaux protegés de l'Afrique Noire*. Dakar.

Devos, A. 1968. The need for nature reserves in East Africa. *Biol. Cons.* 1. (r.t.)

– 1969. Ecological conditions affecting the production of wild herbivorous mammals on grasslands. *Advan. Ecol. Res.* 6. (r.t.)

Dietrich, W.O. 1941. Die säugetierpaläontologischen Ergebnisse der Kohl-Larsen'schen Expedition 1937–1939 im nördlichen Deutsch-Ostafrika. *Zentbl. Miner. Geol. Paläont.*, Ser. B, no.8:217–23. (pal.)

- 1942. Altestquartäre Saugetiere aus der südlichen Serengeti, Deutsch Ost-Africa. *Palaeontogr. Abt. A*, 94:43–133. (pal.)

Dollman, J. G. 1921. *Catalogue of the Selous Collection of Big Game in the British Museum (Natural History)*, pp. 13–14. London: Longmans.

Dorst, J. (ed.). 1958. Investigation upon the present status of ungulates in Africa south of the Sahara. *Mammalia*, 22:357–503.

Dorst, J., and P. Dandelot. 1970. *A Field Guide to the Larger Mammals of Africa*, pp. 274–7, London: Collins.

Dowsett, R. J. 1966. Behaviour and population structure of hartebeest in the Kafue National Park. *Puku*, no.4:147–54. (r.t.)

Duvernoy, G. L. 1851. Note sur une éspèce de buffle fossile, *Bubalus (Arni) antiquus*, decouverte en Algérie. *C. R. Hebd. Séanc. Acad. Sci.*, 33:595–7. (pal.)

Edmond-Blanc, Fr. 1932. En mission dans l'Oubangui-Chari. *Terre Vie*, 2:699.

Eisenberg, J. F., and M. Lockhart. 1972. An ecological reconnaissance of Wilpattu National Park, Ceylon. *Smithsonian Contributions to Zoology*, 101. Washington, D.C.: Smithsonian Institute.

Eltringham, S. K., and M. H. Woodford. 1973. The numbers and distribution of buffalo in the Ruwenzori National Park, Uganda. *E. Afr. Wildl. J.*, 11:151–64.

Emiliani, C. 1955. Pleistocene temperatures. *J. Geol.*, 63:538–78.(cli.)

- 1958. Paleotemperature analysis of core 280 and Pleistocene correlations. *J. Geol.*, 66:264–75.(cli.)

Estes, R. D. 1969. Territorial behaviour of the wildebeest (*Connochaetes taurinus* Burchell, 1823). *Z. Tierpsychol.* 23(3):284–370. (r.t.)

Estes, R. D., and J. Goddard. 1967. Prey selection and hunting behavior of the African wild dog. *J. Wildl. Mgmt.* 31(1):52–70. (r.t.)

Fairall, N. 1968. The reproductive seasons of some mammals in the Kruger National Park. *Zool. Afr.*, 3(2):189–210.

Field, C. R. 1968. A comparative study of the food habits of some wild ungulates in the Queen Elizabeth Park, Uganda: preliminary report. *Symp. Zool. Soc. (London)*, 21:135–51.

Finlayson, R. 1965. Spontaneous arterial disease in exotic animals. *J. Zool. (London)*, 147:239–343.

Fitzsimons, F. W. 1920. *The Natural History of South Africa*, vol. 3, pp.141–9. London: Longmans.

Flint, R. F., and E. S. Deevey. 1957. Postglacial hypsithermal interval. *Science*, 125(3240):182–4. (cli.)

Flower, S. S. 1900. Notes on the fauna of the White Nile and its tributaries, II: mammals. *Proc. Zool. Soc. (London)*, 952–5.

- 1931. Contributions to our knowledge of the duration of life in vertebrate animals, 5:mammals, *Proc. Zool. Soc. (London)*, 145–234.

Foa, E. 1899. *After Big Game in Central Africa*, trans. from French. London: Black.

Foster, J. B., and M. J. Coe. 1968. The biomass of game animals in Nairobi National Park, 1960–66. *J. Zool. (London)*, 155:413–25.

Fraser, A. F. 1968. *Reproductive Behaviour in Ungulates*. London: Academic Press. (r.t.)

- 1974. *Farm Animal Behavior*. London: Bailliere Tindall. (r.t.)

Frèhkop, Z. S. 1943. Mammifères. *Exploration du Parc National Albert: Mission Frechkop (1937–1938)*, fasc. 1, pp. 122–23. Brussels: Institut des Parcs Nationaux Congo Belge.

Funaioli, U. 1971. *Guida breve dei mammiferi della Somalia*. Instituto Agronomico per l'Oltremare–Biblioteca Agraria Tropicale.

Gaillard, C. 1934. Contribution a l'étude de la faune préhistorique de l'Egypte. *Arch. Mus. Hist. Nat. (Lyon)*, 14(3):1–126. (pal.)

Gasthuys, P. 1930. Essais de capture et de domestication d'animaux sauvages. *Bull. Agr. Congo Belge*, 21(2):382.

Gentry, A. W. 1967. *Pelorovis oldowayensis* Reck, an extinct bovid from East Africa. *Brit. Mus. Nat. Hist., Fossil mammals of Africa*, no.22:243–99. (pal.)

– 1970. The Bovidae (Mammalia) of the Fort Ternan fossil fauna. In *Fossil Vertebrates of Africa*, L. S. B. Leakey and R. J. C. Savage (eds.), vol. 2, pp.243–323. London: Academic Press. (pal.)

– 1974. A new genus and species of Pliocene boselaphine (Bovidae, Mammalia) from South Africa. *Ann. S. Afr. Mus.*, 65:145–88. (pal.)

– 1978. Bovidae. *Evoluton of African Mammals*, In V. J. Maglio and H. B. S. Cooke (eds.), pp. 540–72. Cambridge; Harvard University Press. (pal.)

– 1979. Fossil Bovidae (Mammalia) of Langenbaanweg, South Africa. *Ann. S. Afr. Mus.* (pal.)

Glover, P. E. 1968. The role of fire and other influences on savannah habitat with suggestion for further research. *E. Afr. Wildl. J.*, 6. (r.t.)

Glover, P. E., J. Glover, and M. E. Gwynne. 1962. Light rainfall and plant survival in dry grassland vegetation. *J. Ecol.*, 50. (r.t.)

Goodwin, L. G. 1972. Scientific report: Zoological Society: Nuffield Institute of Comparative Medicine. *J. Zool.* (London), 166:590.

Graaf, G. de, K. C. A. Schultz, and P. T. van der Walt. 1973. Notes on rumen contents of Cape buffalo *Syncerus caffer* in the Addo Elephant National Park. *Koedoe*, no.16:45–58.

Graham, P. 1967. An analysis of the numbers of game and other large mammals killed in tsetse fly control operations in northern Bechuanaland 1942 to 1963. *Mammalia*, 31:186–204.

Gray. 1837. In Roberts (1951).

– 1843. In Roberts (1951).

Gray, J. E. 1821. On the natural arrangement of vertebrate animals. *London Med. Repository*, 15:296–310.

– 1872. *Catalogue of Ruminant Mammalia (Pecora, Linnaeus) in the British Museum*. 8 vols. London.

– B. pumilus. *Ann. Nat. Hist.* (4) 12:499–500.

– B. pumilus. *Ann. Nat. Hist.* (4) 13:258–9.

Grimsdell, J. J. R. 1969. Ecology of the Buffalo in western Uganda. Ph.D. Thesis, Cambridge University.

– 1973a. Age determination of the African buffalo, *Syncerus caffer* Sparrman. *E. Afr. Wildl. J.*, 11:31–54.

– 1973b. Reproduction in the African buffalo, *Syncerus caffer*, in western Uganda. *J. Reprod. Fert., Suppl.* 19:301–16.

Grimwood, I. R. C., C. W. Benson, and W. F. H. Ansell. 1958. The present day status of ungulates in Northern Rhodesia. *Mammalia*, 22:451–67.

Gromier, D. 1949. *Grands fauves d'Afrique*, vol. 3, pp. 97–187. Paris.

Gross, H. 1958. Die bisherigen Ergebnisse von C-14 Messungen und Paläontologischen Untersuchungen fur die Gliederung und Chronologie des Jung-Pleistozans in Mitteleuropa und den Nachbargebieten. *Eisz. Gegenw.*, 9:155–87. (cli.)

Groves, C. P. 1975. Notes on the gazelles, I: *Gazella rufifrons* and the zoogeography of Central African Bovidae. *Z. Säuget.*, 40:308–19.

Grubb, P. 1972. Variation and incipient speciation in the African buffalo. *Z. Säuget.*, 37:121–44.

Grzimek, B. 1970. *Among Animals of Africa*. London: Collins.

Guggisberg, C. A. W. 1961. *Simba, the Life of the Lion*. Cape Town: Howard Timmins. (r.t.)

– 1963. *Game Animals of Eastern Africa*, 8th ed. Nairobi: Patwa News Agency.

Guilbride, P. D. L. et al. 1963. Tuberculosis in the free-living African (Cape) buffalo (*Syncerus caffer caffer* Sparrman). *J. Comp. Pathol.*, 73:337–48.

236 *References*

van Gylswyk, N. O., and D. Giesecke. 1973. A summary of preliminary findings in a rumen microbiological investigation on wild ruminants. *Koedoe*, 16:191–4. (r.t.)

Hafez, E. S. E. (ed.). 1975. *The Behaviour of Domestic Animals*. 3rd ed. London:Bailliere Tindall. (r.t.)

Hall, P. 1976. Priorities for wildlife conservation in northeastern Nigeria. *Nigerian Field*, 41(3):99–112.

Haltenorth, T. 1953. *Hornträger*, vol. 1. Stuttgart: Kosmos.

Hamilton, W. R. 1973. The Lower Miocene ruminants of Gebel Zelten, Libya. *Bull. Brit. Mus. Nat. Hist. (Geol.)*, 21:50–73. (pal.)

Hancock, J. 1953. Grazing behaviour of cattle. *Animal Breed. Abstr.*, 21:1–13. (r.t.)

Hanks, J. 1969. Techniques for marking large African mammals. *Puku*, no.5:65–86.

Happold, D. C. D. 1973. The distribution of large mammals in West Africa. *Mammalia*, 37:88–93.

Harris, J. M. 1976. Bovidae from the East Rudolf Succession. In *Earliest Man and Environments in the Lake Rudolf Basin*, Y. Coppens, F. C. Howell, G. L. Isaac, and R. E. F. Leakey (eds.), pp. 293–301. Chicago: University of Chicago Press. (pal.)

Harthoorn, A. M. 1958. Comparison of food intake and growth-rate of the African buffalo (*Syncerus caffer*) with indigenous cattle: preliminary report. *Veterin. Rep.*, 70(46):939–40.

Heady, H. F. 1966. Influence of grazing on the composition of *Themeda triandra* grassland, East Africa. *J. Ecol.*, 54. (r.t.)

Heck, H., D. Wurster, and K. Benirschke. 1968. Chromosome study of members of the subfamilies Caprinae and Bovinae, family Bovidae; the musk ox, ibex, aoudad, Congo buffalo, and gaur. *Z. Säuget.*, 33:172–9.

Hedberg, I., and O. Hedberg. 1968. Conservation of vegetation in Africa south of the Sahara. *Acta Phytogeogr. Suec.*, 54. (r.t.)

Hedger, R. S. 1972. Foot-and-mouth disease and the African buffalo (*Syncerus caffer*). *J. Comp. Pathol.*, 82:19–28.

Hedger, R. S., A. J. Forman, and M. H. Woodford. 1973. Le virus de la fièvre apheteuse chez le buffle est-africain. *Bull. Epizootic Dis. Afr.*, 21:101–3.

Hendey, Q. B. 1968. The Melkbos site: and Upper Pleistocene fossil occurrence in the southwestern Cape Province. *Ann. S. Afr. Mus.*, 52:89–119. (pal.)

– 1970. A review of the geology and palaeontology of the Plio-Pleistocene deposits at Langebaanweg, Cape Privince. *Ann. S. Afr. Mus.*, 56:75–117. (pal.)

Henshaw, J., and C. Greeling. 1973. A description of the savanna buffalo of Yankeri Game Reserve, north-eastern Nigeria. *Mammalia*, 37:94–100.

Herrig, D. M., and A. O. Haugen. 1969. Bull bison behavior traits. *Proc. Iowa Acad. Sci.*, 76:245–62. (r.t.)

Hewitt, J. 1931. *A Guide to the Vertebrate Fauna of the Southern Cape Province*, part 1. Cape Town.

Hill, J. E., and T. D. Carter. 1941. The mammals of Angola, Africa. *Bull. Amer. Mus. Nat. Hist.*, 78:162.

Hodgson. 1847. *J. Asiatic Soc. Bengal*, nas. 16 (7):709.

Hofmann, R. R., and D. R. M. Stewart. 1972. Grazer or browser: a classification based on the stomach structure and feeding habits of East African ruminants. *Mammalia*, 36:226–40.

Höhnel, L. von. 1894. *Discovery of lakes Rudolf and Stefanie*, 2 vols., trans. London: Longmans.

Hoier, R. 1952. *Mammifères du Parc National Albert*. Brussels: Institut des Parcs Nationaux du Congo Belge.

Holub, E. 1881. *Seven years in South Africa: Travel, Researches, and Hunting Adventures, between the Diamond fields and the Zambesi (1872–1879)*, 2 vols. London.

Hopwood, A. T. 1936. New and little-known fossil mammals from the Pleistocene of Kenya Colony and Tanganyika Territory. *Ann. Mag. Nat. Hist.* (London), 17:636–41. (pal.)

Hopwood, A. T., and J. P. Hollyfield. 1954. An annotated bibliography of the fossil mammals of Africa (1742–1950). *Brit. Mus. Nat. Hist., Fossil Mammals of Africa*, no.8:1–194. (pal.)

de la Hunt, T. E. 1954. r.t. The value of browse shrubs and bushes in the Lowveld of the Gwanda area, S. Rhodesia. *Rhodesia Agri. J.* 51(4):251–62.

Hunter, R. F., and G. E. Davies. 1963. The effect of method of rearing on the social behaviour of Scottish blackface hoggets. *Animal Prod.*, 5:183–94. (r.t.)

Huntley, B. J. 1978. Ecosystem conservation in Southern Africa. In *Biogeography and Ecology of Southern Africa*, M. J. A. Werger (ed.), p. 1364. The Hague: Dr. W. Junk bv Publishers.

Ittner, N. R., C. F. Kelly, and H. R. Guilbert. 1951. Water consumption of Hereford and Brahman cattle and the effect of cooled drinking water in hot climates. *J. Anim. Sci.*, 10:742–51.

(I. M.) 1950. Les buffles d'Afrique. *Rev. Coloniale Belge*, no. 104:87–89, 107:192–94.

Jackson, F. J. et al. 1894. *Big Game Shooting*, 2 vols. London: Badminton Library.

– 1899. In *Great and Small Game of Africa*, H. A. Bryden (ed.). London: Rowland Ward.

Jarman, P. J. 1974. The social organization of antelope in relation to their ecology. *Behaviour*, 48:215–67. (r.t.)

Jeannin, A. 1936. *Mammifères sauvages du Cameroun*. Paris.

– 1951. *La faune africaine*. Paris: Payot.

Jeannin, A., and M. Barthe. 1958. L'evolution africaine et la persistance de la faune sauvage. *Mammalia*, 22:328–35. (r.t.)

Jobaert, A. J. 1952. Le buffle de Cafrerie ou buffle noir sud du Congo Belge. *Zooleo*, 17.

Johnson, Sir H. 1906. *Liberia*, 2 vols. pp. 370, 730–5. London: Hutchinson.

Johnson, H. H. 1885. General observations on the fauna of Kilimanjaro. *Proc. Zool. Soc.* (London), 214–18.

– 1897. *British Central Africa*. London: Methuen.

Kaliner, G., and C. Staak. 1973. A case of orchitis caused by *Brucella abortus* in the African buffalo. *J. Wildl. Dis.*, 9:251–3.

Kenneth, J. H. Gestation periods. *Tech. Comm. C. A. B.*

Kettlitz, W. K. 1962. The distribution of some of the larger game mammals in the Transvaal (excluding the Kruger National Park). *Ann. Cape Prov. Mus.*, 2:118–37.

Kieffer, Ch. 1953. Les reserves de faune du Cameroun. *Mammalia*, 17:270–74.

King, J. M., and B. R. Heath. 1975. Experiences at the Galana Ranch. *World Animal Rev.*, 16;23–30.

Kirk, J. 1864. List of Mammalia met with in Zambesia, East Tropical Africa. *Proc. Zool. Soc.* (London), 648–60.

Koller, O. 1935. Die Rassen des Westafrikanischen Rotbuffels. *Sitzungsberichte Mathem.-Naturw.*, 1(144):419–53.

Krumbiegel, L. 1933. Wie füttere ich gefangene Tiere? Leipzig: 17.

Kruuk, H. 1972. *The Spotted Hyena*. Chicago: University of Chicago Press.

Lake, A. 1953. *Killers in Africa*, pp. 145–62. New York: Doubleday.

Lamotte, M. 1942. La faune mammalogique du Mont Nimba (Haute Guinée). *Mammalia*, 6:114–19.

Lamprey, H. F. 1963. Ecological separation of the large mammal species in the Tarangire Game Reserve, Tanganyika. *E. Afr. Wildl. J.*, 1:63–92.

Lavauden, L. 1927. Contribution à l'histoire naturelle des buffles. *Rev. franc. mammal.*, 1:10–39.

– 1934. *Les grands animaux de chasse de l'Afrique française*. Paris.

238 References

Leakey, R. E. F. 1969. Early *Homo sapiens* remains from the Omo River region of southwest Ethiopia. *Nature*, 222:1132-3. (pal.)

Ledger, H. P. 1963. Weights of some East African mammals. *E. Afr. Wildl. J.*, 1:123-24.

Leroi-Gourhan, A. 1958. Resultats de l'analyse pollinique du gisement d'El Guettar (Tunisie). *Bull. S.P.F.*, 55(9). (cli.)

Leuthold, W. 1972. Home range, movements, and food of a buffalo herd in Tsavo National Park. *E. Afr. Wildl. J.*, 10:237-43.

Lichtenstein, H. 1930. *Travels in Southern Africa*, vol. 2, trans. A. Plumtree.

Linear, M. 1976. *Buffalo and Antelope Ranching in West Africa*. Prelimany project proposal on behalf of Operation Save the Apes. Munich, mimeographed.

Littlejohn, K. G. 1938. Field notes on bushcow (*Syncerus caffer* subsp.). *Nigerian Field*, 7(1):17-20.

Livingstone, D. 1857. *Missionary Travels and Research in South Africa*. London: John Murray.

- In *Livingstone's Private Journals*, In I. Schapera (ed.). London: Chatto & Windus/Berkeley, University of California Press.

Livingstone, D., and C. Livingstone. 1865. *Narrative of an Expedition to the Zambezi and Its Tributaries*. London: John Murray.

Lönnberg, E. 1933. Description of a fossil buffalo from East Africa. *Ark. Zool.*, 25A, 17:1-32. (pal.)

Lorenz, K. 1966. *On Aggression*. New York: Harcourt Brace Jovanovich. (r.t.)

Loveridge, A. 1921. Notes on East African mammals, collected 1920-1923: Ungulata Bovidae. *Proc. Zool. Soc.* (London), 732-737.

Ludbrook, J. V. 1963. Desertion of a buffalo calf. *Puku*, no.1:216.

Lydekker, R. 1898. *Wild Oxen, Sheep, and Goats of All Lands Living and Extinct*. London.

- 1904. On a buffalo skull from East Central Africa. *Proc. Zool. Soc.* (London), 2:163-5.

- 1906. Description of two mammals from the Ituri Forest (with supplementary note on the buffalo of the Semliki district). *Proc. Zool. Soc.* (London), 2:992-6.

- 1908. *The Game Animals of Africa*. London.

- 1910. On three African buffaloes. *Proc. Zool. Soc.* (London), 2:992-8.

- 1912. *The Ox and Its Kindred*, p. 176. London.

- 1913. The dwarf buffalo of southern Nigeria; with a revision of the dwarf buffaloes of western Africa. *Proc. Zool. Soc.* (London), 1:234-41.

- 1926. *The Game Animals of Africa*, 2nd ed., rev. J. G. Dollman, pp. 54-72. London: Rowland Ward.

Maarveld, G. C., and Th. van der Hammen. 1959. The correlation between Upper Pleistocene pluvial and glacial stages. *Geol. Mijnbouw.*, n.s. 21:40-45. (cli.)

Maclatchy, A. R. 1932. Les buffles du Gabon. *Terre Vie*, 2(10):584-96.

McBurney, C. B. M. 1960. *The Stone Age of Northern Africa*. Great Britian: Pelican. (cli.)

McConnell, E. E., R. C. Tustin, and V. de Vos. 1972. Anthrax in an African buffalo (*Syncerus caffer*) in the Kruger National Park. *J. S. Afr. Vet. Med. Ass.*, 43:181-7.

McDairmid, A. 1960. *Diseases of Free-living Wild Animals*. F.A.O. Animal Health Branch Monograph No. 1. Rome.

McHugh, T. 1958. Social behavior of the American buffalo. *Zoologica*, 43:1-40. (r.t.)

McMahan, C. A. 1964. Comparative food habits of deer and three classes of livestock. *J. Wildl. Mgmt.* 28(4): (r.t.)

Malbrant, R. 1935. Note au sujet de la classification des buffles africains. *Bull. Mus. Hist. Nat.*(Paris), 2me ser., 7.

Malterre, J. 1947. *Compte rendu du XIII⁰ Congres International de Zoologie*. Paris.

Markus, J. 1948. Data on duration of pregnancy, birthweight, and growth of the buffalo. *Animal-Breed. Abstr.*, March.

Marsboom, R. 1950. De buffel en zijn economische betekenis voor Belgisch-Congo. *Bull. Agric. Congo Belge*, 41(3):773-92.

Maydon, H. C. 1957 (reprint). In *Big Game Shooting in Africa*. London: Seeley, Service.

Meinertzhagen, R. 1938. Some weights and measurements of large mammals. *Proc. Zool. Soc.* (London), ser. A, 108:433–9.

Mentis, M. T. 1970. Estimates of natural biomasses of large herbivores in the Umfolozi Game Reserve area. *Mammalia*, 34:363–93.

Mitchell, B. L. 1963. A first list of plants collected in the Kafue National Park. *Puku*, no.1:75–191. (r.t.)

Mitchell, B. L., J. B. Shenton, and J. C. N. Uys. 1965. Predation on large mammals in the Kafue National Park, Zambia. *Zool. Africana*, 1(2):297–319.

Mloszewski, M. J. 1974. Notes on the buffalo. *Black Lechwe*, 11(3):

Mohr, C. O., and W. A. Stumf. 1966. Comparison of methods for calculating areas of animal activity. *J. Wildl. Mgmt.*, 30(2): (r.t.)

Monard. A. 1935. *Contribution à la mammalogie d'Angola et prodrome d'une faune d'Angola*, vol. 6, pp. 264–6. Lisbon; Arquivos do Museu Bocage.

– 1951. Resultats de la Mission Zoologique Suisse au Cameroun, 2: Mammiferes. *Memoires de l'Institut Français d'Afrique Noire (Centre du Cameroun)*, 42.

Monod, T. 1963. The Late Tertiary and Pleistocene in the Sahara. In *African Ecology and Human Evolution*, F. C. Howell and F. Bourlière (eds.) pp. 117–29. Chicago: Aldine. (cli.)

Moreau, R. E., G. H. E. Hopkins, and R. W. Hayman. 1945/46. The type localities of some African mammals. *Proc. Zool. Soc.* (London), 115:387–447.

Moreau, R. E. 1963. The distribution of tropical African birds as an indicator of past climatic changes. In *African Ecology and Human Evolution*, F. C. Howell and F. Bourlière (eds.), pp. 28–42. Chicago: Aldine. (cli.)

Moss, C. 1976. *Portraits in the Wild*. London: Hamish Hamilton.

Müller-Schwartze, D., and C. Müller-Schwartze. 1970. Lip-smacking in the pronghorn (*Antilocapra americana*). *Z. Säuget.*, 35:353–6. (r.t.)

Myers, N. 1972. *The Long African Day*. New York: Macmillan/London: Collier Macmillan.

Neitz, W. O. 1965. A checklist and hostlist of the zoonoses occurring in mammals and birds in South and South West Africa. *Onderspoort J. Vet. Res.*, 32(2);189–374. (r.t.)

Neumann, A. H. 1898. *Elephant Hunting in East Equatorial Africa*. London: Rowland Ward.

Nilsson, E. 1964. Pluvial lakes and glaciers in East Africa. *Stockholm Contrib. Geol.*, 11:21–57. (cli.)

Norton-Griffiths, M. 1975. The numbers and distribution of large mammals in Ruaha National Park, Tanzania. *E. Afr. Wildl. J.*, 13:121–40.

Nosti, J. 1950. Bufalos fernandinos. *Agricultura* (Madrid), no. 220

Parkyns, M. 1853. *Life in Abyssinia*, 2 vols. London: John Murray.

Parsons, L. M. 1966. *A Commentary and Summary of Available Published Information on Cow Behaviour*. London: Min. Agric., Land Service, mimeographed. (r.t.)

Pechuel-Loesche, 1888 Afrikanische Buffel. Giessen, Spengel, *Zool. Jahrb.*, iii, *Ab. f. Systemt.* :704–24.

Peel, C. V. A. 1900. *Somaliland*, p. 290. London: Robinson.

Pennant, T. 1771. *Synopsis of Quadrupeds*. Chester.

– 1781. *History of Quadrupeds*, 2 vols. London.

Penrice, G. W. 1899. In *Great and Small Game of Africa*, H. A. Bryden (ed.). London: Rowland Ward.

Peters, W. C. H. 1869. Säugetiere gesammelt von Baron C. C. von der Decken auf zeinen Reisen im A equatorialen Ostafrica. In C. C. von der Decken, *Reisen in Ost-Africa*, O. Kersten (ed.), 8 vols., appendix. Leipzig.

Petrides, G. A., and W. G. Swank. 1965. Population densities and the range-carrying capacity for large mammals in Queen Elizabeth National Park, Uganda. *Zoologica Africana*, 1(1):209–25.

Petrocchi, C. 1956. I *Leptobos* di Sahabi. *Boll. Soc. Geol. Ital.*, 75(1):206–38. (pal.)

Philips, J. 1965. Fire as a master and servant: its influence on bioclimatic regions of trans-Sahara Africa. *Fourth Annual Tall Timber Conference.* (r.t.)

Pienaar, U. de V. 1961. A second outbreak of anthrax amongst game animals in the Kruger National Park. *Koedoe,* 4:4–18.

– 1963. The large mammals of the Kruger National Park: their distribution and present-day status. *Koedoe,* 6:1–38.

– 1968. The use of fire as a tool in wildlife management in the Kruger National Park. In *A Practical Guide to the Study of Larger Herbivores,* F. B. Golley and H. K. Buechner (eds.). I.B.P. Handbook 7. (r.t.)

– 1969a. Observations on developmental biology, growth, and some aspects of the population ecology of African buffalo (*Syncerus caffer caffer* Sparrman) in the Kruger National Park. *Koedoe,* 12:29–52.

– 1969b. Predator–prey relationships amongst the larger mammals of the Kruger National Park. *Koedoe,* 12:108–76.

Pienaar, U. de V., P. W. van Wyk, and N. Fairall. 1966. An aerial census of elephant and buffalo in the Kruger National Park and the implications thereof on intended management schemes. *Koedoe,* 9:40–108.

Pitman, C. R. S. 1934. *A Report on a Faunal Survey of Northern Rhodesia.* Livingstone: Government Printer.

– 1942. *A Game Warden Takes Stock.* London: Nisbit.

Plowes, D. C. H. 1957. The seasonal variation of crude protein in twenty common veld grasses at Matopos, Southern Rhodesia, and related observations. *Salisbury, Rhodesia Agr. J.,* 54(1):33–55. (r.t.)

Poche, R. M. 1976. A checklist of the mammals of National Park W, Niger, West Africa. *Nigerian Field,* 41 (3):113–15.

Pockock, R. I. 1910. On the specialised cutaneous glands of ruminants. *Proc. Zool. Soc.* (London), 2:840–986.

Pomel, A. 1893. *Bubalus antiquus.* Carte Geol. Alger. Paleont. Monogr. (pal.)

Pons, A., and P. Quezel. 1957. *Premiere étude palynologique de quelques paleosols sahariens.* Travaux de l'Inst. de Recherches Sahar. 16. Paris. (cli.)

– 1958. Palynologie: Premieres rémarques sur l'etude palynologique d'un guano fossil du Hoggar. *C. R. Acad. Sci.,* 244:2290–2. (cli.)

Pratt, D. J., P. J. Greenway, and M. D. Gwynne. 1966. A classification of East African rangeland. *J. Appl. Ecol.,* 3. (r.t.)

Prior, H. 1944. Africa's ugliest beast. *Field & Stream,* 49(5):9–11, 57, 49(6):32–33.

Prunier, R. 1937. Conference donnée le 17 decembre 1936 a la Societe. *Bull. Soc. Nat. d'Acclim.* (Paris).

Rattray, J. M. 1957. The grasses and grass associations of Southern Rhodesia. *Salisbury, Rhodesia Agr. J.,* 54(3):197–234. (r.t.)

Rattray, J. M., and H. Wild. 1955. Report on the vegetation of the alluvial basin of the Sabi valley and adjacent areas. *Salisbury, Rhodesia Agr. J.,* 52(6):484–501. (r.t.)

Reboussin, R. 1953. *Contes de ma vie sauvage,* pp. 213–14. Paris.

Reck, H. 1928. *Pelorovis oldowayensis* n.g. n.sp. *Wiss. Ergebn. Oldoway-Exped. 1913* (N.F.), 3:57–67. (pal.)

Renshaw, G. 1904. Notes on June 7th meeting. *Proc. Zool. Soc.* (London), 2:131.

Roberts, A. 1948. The Cape buffalo. *Afr. Wild Life,* 2(1):39–49.

– 1951. *The Mammals of South Africa.* Cape Town; Central News Agency.

Robinette, W. L. 1963. Weights of some of the larger mammals of Northern Rhodesia. *Puku,* no.1:207–15.

Rochebrune, A. T. de. Sur une espèce nouvelle du genre *Bubalus* provenant de la haute Sénégambie. *Bull. Soc. Philom.* (Paris), 9:15–19.

Rockwell, R. R. 1934. On the trail of the African buffalo. *Nat. Hist.* (New York), 34(1).

Rode, P. 1943. Mammifères ongules de l'Afrique Noire, I: Bovides. Paris: Larousse.

Roe, F. G. 1970. *The North American Buffalo: A Critical Study of the Species in Its Wild State*, 2nd ed. Toronto: University of Toronto Press. (r.t.)

Roosevelt, T., and E. Heller. 1914. *Life Histories of African Game Animals*, 2 vols.

Rosevear, D. R. 1951. *Nigerian Mammals*, pp. 22–23. Lagos: Government Printer.

– 1953. *Checklist and Atlas of Nigerian Mammals*, p. 123 Lagos: Government Printer.

Roure, G. 1962. *Animaux sauvages de Côte d'Ivoire*. Abidjan: Ministrie de l'Agriculture et de la Cooperation.

Schaller, G. B. 1972. *The Serengeti Lion: A Study of Predator–Prey Relationships*. Chicago: University of Chicago Press.

Schlegel, H. 1880. *Muséum d'Histoire Naturelle des Pays-Bas. Revue méthodique et critique des collections déposées dans cet établissement*. Leyden:I–VIII.

– 188? On the zoological researches in West Africa. *Notes Leyd. Mus.*, 3:53–58. (Note xiv.)

Scholoeth, R. 1958. Das Scharren bei Rind und Pferd. *Z. Säuget.*, 23:139–48. (r.t.)

Schouteden, H. 1947. De zoogdieren van Belgisch-Congo en van Ruanda-Urundi. *Annalen van het Museum van Belgisch-Congo, C. Dierkunde*, Reeks II, Deel III, Afl. 1–3.

Schwartz, E. 1920. Huftiere aus West- und Zentralafrica, Ergebnisse der zweiten Deutschen Zentral-Africa-Expedition, 1910–1911. *Zoologie*, June.

Sclater, P. L. 1864. On the mammals collected and observed by Capt. J. H. Speke during the East African expedition. *Proc. Zool. Soc.* (London), 98–106.

– 1898. Notes on exhibition of 3 heads from the Gambia. *Proc. Zool. Soc.* (London),

Sclater, W. L. 1900. *The Mammals of South Africa*, 2 vols. London.

Seeley, H. G. Pleistocene of South Africa. *Geol. Mag.* 8(3):199–202. (cli.)

Selous, F. C. 1881. *A Hunter's Wanderings in Africa*. London: Bentley.

– 1899. In *Great and Small Game of Africa*, H. A. Bryden (ed.). London: Rowland Ward.

– 1908. *African Nature Notes and Reminiscences*. London: Macmillan.

Sheppe, W., and P. Haas. 1976. Large mammal populations of the lower Chobe river, Botswana. *Mammalia*, 40:223–43.

Shortridge, C. C. 1934. *Mammals of South West Africa*, vol. 2, pp. 439–48. London: Heinemann.

Sidney, O. J. 1965. *The Distribution of the Buffalo Today and Fifty Years Ago: Fact-finding Survey of the African Fauna*, part 6. London: British Museum (Natural History), mimeograph. Also:*Trans. Zool. Soc.* (London), 30:1–396.

Sinclair, A. R. E. 1970. Studies of the ecology of the East African buffalo. Ph.D. thesis, Oxford University.

– 1972. Long term monitoring of mammal populations in the Serengeti: census of non-migratory ungulates, 1971. *E. Afr. Wildl. J.*, 10:287–97.

– 1973a. Population increases of buffalo and wildebeest in the Serengeti. *E. Afr. Wildl. J.*, 11:93–107.

– 1973b. Regulation and population models for a tropical ruminant. *E. Afr. Wildl. J.*, 11:307–16.

– 1974a. The natural regulation of buffalo population in East Africa, I: Introduction and resource requirements. *E. Afr. Wildl. J.*, 12:135–54.

– 1974b. The natural regulation of buffalo population in East Africa, II: Reproduction, recruitment, and growth. *E. Afr. Wildl. J.*, 12:169–83.

– 1974c. The natural regulation of buffalo population in East Africa, III: Population trends and mortality. *E. Afr. Wildl. J.*, 12:185–200.

– 1974d. The natural regulation of buffalo population in East Africa, IV: The food supply as a regulating factor, and competition. *E. Afr. Wildl. J.*, 12:291–311.

– 1974e. The social organization of the East African buffalo (*Syncerus caffer* Sparrman). In *The Behaviour of Ungulates and Its Relation to Management*, V. Geist and F. Walther (eds.), n.s. no. 24, pp. 676–89, Morges, Switzerland: I.U.C.N. Publications.

- 1975. The resource limitation of trophic levels in tropical grassland ecosystems. *J. Animal Ecol.*, 44:497–520.
- 1977. *The African buffalo: A study of resource limitation of populations.* Chicago: University of Chicago Press.

Sinclair, A. R. E., and P. Duncan. 1972. Indices of condition in tropical ruminants. *E. Afr. Wildl. J.*, 10:143–9.

Sinclair, A. R. E., and M. D. Gwynne. 1972. Food selection and competition in the East African buffalo (*Syncerus caffer* Sparrman). *E. Afr. Wildl. J.*, 10:77–89.

Smith, A. 1834. In Roberts (1951).

Smith, A. H. 1939. Measurements and descriptions of bushcow. Nigerian Field, 8(1):39–42.

Smith, H. 1827. In Roberts (1951).

- 1850.In Roberts (1951).

Smith, M., and G. R. Price. 1973. The logic of animal conflict. *Nature*, 246:(r.t.)

Smithers, R. H. N. 1966. *The Mammals of Rhodesia, Zambia, and Malawi*, pp. 133–4. London: Collins.

Smuts. 1832. In Roberts (1951).

Sparrman, A. 1779. A reference to the African buffalo. *Kongl. Svenska Vet.-Akad. Handl.* (Stockholm), 40:79

Sparrman, A. 1785. *A Voyage to the Cape of Good Hope Towards the Antarctic Polar Circle and Round the World but Chiefly into the Country of the Hottentots and Caffres, from the years 1772–1776*, 2 vols., trans. G. Foster. Perth: Morison.

Spencer, R. 1945. Experiences with buffalo in Burma and Rhodesia. *J. Bombay Nat. Hist. Soc.*, 45(2):232–3.

Spinage, C. A. 1971. Two records of pathological conditions in the impala (*Aepyceros melampus*). *J. Zool.* (London), 164:269. (Description of a pathological condition in buffalo calf tooth development.)

Stevenson-Hamilton, J. 1912. *Animal Life in Africa.* New York: Dutton.

- 1925. Annual report to the National Parks Board of Trustees. S. Africa, mimeo.
- 1929. The Low-Veld: Its wildlife and its people.

Stewart, D. R. M., and J. Stewart. 1963. The distribution of some large mammals in Kenya. *J. E. Afr. Nat. Hist. Soc.*, 24(3):45.

- 1970. Food preference data by faecal analysis for African plains ungulates. *Zool. Africana*, 5.

Stigand, C. H., and D. D. Lyell. 1906. *Central African Game and Its Spoor*, pp. 82–85. London: Cox.

Stockley, C. H. 1948. *African Camera Hunts.* London: Country Life.

Strong, R. P. (ed.). 1930. *The African Republic of Liberia and the Belgian Congo: Harvard Afrcian Expedition (1926–1927)*, vol. 2. pp. 614–15. Cambridge: Harvard University Press.

Swayne, H. G. C. 1900. *Seventeen Trips through Somaliland and a Visit to Abyssinia.* London: Rowland Ward.

Sweeney, R. C. H. 1959. A check list of the mammals of Nyasaland. Blantyre: Nyasaland Society.

Swynnerton, G. H. 1945/46. A revision of the type localities of mammals occurring in the Tanganyika Territory. *Proc. Zool. Soc.* (London), 115:49–84.

Talbot, L. M., and D. R. M. Stewart. 1964. First wildlife census of the entire Serengeti–Mara region, East Africa. *J. Wildl. Mgmt.*, 28(4).

Talbot, L. M., and J. S. G. McCulloch. 1965. Weight estimates for East African mammals from body measurements. *J. Wildl. Mgmt.*, 29(1).

Temminck, C. J. 1853. *Esquisses zoologiques sur la Cote de Guine*, part 1, *Les mammiferes*, pp. 239–41. Leiden: Brill.

Tener, J. S. 1954. *A Preliminary Study of the Musk-Oxen of Fosheim Peninsula, Ellesmere Island, N.W.T.* Canada Wildlife Service, Wildlife Management Bulletin, 1st ser., no. 9. (r.t.)

Thomas, O. 1885. Report on the mammals obtained and observed by Mr. H. H. Johnson on Mount Kilimanjaro. *Proc. Zool. Soc.* (London), 219–22.

Thunberg. 1811. In Roberts (1951).

Tulloch, D. G. 1978. The water buffalo, *Bubalus bubalis*, in Australia: grouping and home range. *Commonwealth Sci. Ind. Res. Org. Austral. Wildl. Res.*, 5(3): 327–54. (r.t.)

Uganda Game Department Annual Reports. 1925–1960. Government Printer, Kampala.

Vaillant, F. Le 1790. *Travels from the Cape of Good Hope into the Interior Parts of Africa*, no. 12. Koedoe.

Van Campo, M. 1958. Analyse pollinique des depots wurmiens d'El Guettar (Tunisie). *Geobot. Inst. Rübel*, 34:133–5. (cli.)

– 1959. Analyses polliniques dans le sud tunisien. *C.R. IX Congr. Int. Bot.*, Montreal, 2:409–10. (cli.)

Van Der Hammen, Th., and E. Gonzalez. 1960. Upper Pleistocene and Holocene climate and vegetation of the "Sabana de Bogota" (Columbia, S. Am.). *Leidse. Geol. Meded.*, 25:261–315. (cli.)

Van der Schiff, H. P. 1959. Weidingsmoontlikhede en Weidingsprobleme in die Nasionale Krugerwildtuin. *Koedoe*, 2:96–127.

Van Wyk, P., and N. Fairall. 1969. The influence of the African elephant on the vegetation of the Kruger National Park *Koedoe*, 12. (r.t.)

Van zinderen Bakker, E. M. 1958. *Palynology in Africa: Fifth Report Covering the Years 1956, 1957*. Bloemfontein. (cli.)

– 1962. Botanical evidence for Quaternary climates in Africa. *Ann. Cape Prov. Mus.*, 2:16–31. (cli.)

Van Zyl, J. H. M. 1966. Ovulation in the South African buffalo (*Syncerus caffer caffer* Sparrman) in the S. A. Lombard Nature Reserve, Bloemhof, Transvaal. *Fauna Flora* (Pretoria), 17:37.

Verheyen, R. 1951. *Contribution a l'étude ethologique des mammiferes du Parc national de l'Upemba*. Brussels: Institut des Parcs Nationaux Congo Belge, Exploration du Parc National de l'Upemba.

– 1954. Contribution a l'ethologie du buffle noir *Bubalus caffer* (Sparrman). *Mammalia*, 18:

Verschuren, J. 1958. *Ecologie et biologie des grands mammiferes*, fasc. 9, Brussels: Institut des Parcs Nationaux du Congo Belge, Exploration du Parc National de Garamba.

Vesey-Fitzgerald, D. F. 1960. Grazing succession among East African game animals. *J. Mammal.*, 41:161–72.

– 1965. The utilization of natural pastures by wild animals in the Rukwa Valley, Tanganyika. *E. Afr. Wildl. J.*, 3.

– 1970. The origin and distribution of valley grasslands in East Africa. *J. Ecol.*, 58 (r.t.)

Vidler, B. O. et al. 1963. The gestation and parturition of the African buffalo (*Syncerus caffer caffer* Sparrman). *E Afr. Wildl. J.*, 1:122–3.

Vincent, J. 1962. The distribution of ungulates in Natal. *Ann. Cape Prov. Mus.*, 2:110–7. (r.t.)

Walther, F. R. 1964. *Z. Tierpsychol.* 21:871–90. (r.t.)

Ward, R. 1962. *Records of Big Game*, 11th ed. London: Rowland Ward.

Waterhouse, G. 1932. *Simon van der Stel's Journal of his Expedition to Namaqualand, 1685–6*. Dublin:Figgis.

Watson, R. M., and O. Kerfoot. 1964. A short note on the intensity of grazing in the Serengeti Plains by plains game. *Z. Säuget.*, 29. (r.t.)

Weir, J. S. 1969. Chemical properties and occurrence on Kalahari sand of salt licks created by elephants. *J. Zool.* (London), 158:293–310.

References

References

West, O. 1965. *Fire in vegetation and its use in pasture management, with special reference to tropical and subtropical Africa.* Farnham: Commonwealth Agriculture Bureau. (r.t.)

Wild, H., and A. Fernandes (eds.). 1967/68. *Flora Zambesiaca Supplement; Vegetation Map of the Flora Zambesiaca Area.* Salisbury: Collins. (r.t.)

Williams, J. G. 1967. *A Guide to the National Parks of East Africa.* London: Collins.

Wilson, M. 1975. Holocene fossil bison from Wyoming and adjacent areas. M. A. thesis, University of Wyoming, Laramie. (pal.)

Wilson, V. J. 1975. *Mammals of the Wankie National Park, Rhodesia,* pp. 120–21. Salisbury: Trustees of the National Museums and Monuments of Rhodesia.

Wright, B. S. 1960. Predation on big game in East Africa. *J. Wildl. Mgmt.,* 24(1):1–15.

Wright, S. 1922. The effects of inbreeding and cross breeding on guinea pigs, I, II. *U.S.D.A. Bull.,* 1090:1–63. (r.t.)

Woldstedt, P. 1958. Eine neue Kurve der Würm-Eiszeit. *Eisz. Gegenw.,* 9:151–4. (cli.)

Yamba, S. M. 1969. An abandoned buffalo calf. *Puku,* no.5:237.

Zammarano, A. T. 1930. *Le colonie italiane di diretto dominio: fauna e caccia,* pp. 197–9. Ministero dell Colonie.

Zuckerman, S. 1953. The breeding seasons of mammals in captivity. *Proc. Zool. Soc.* (London), 122:827–950.

Zukowsky, L. 1924. Beitrag zur Kenntnis der Säugethiere Deutch-Sudwestafrikas unter besonderer Berucksichtigung des Brobwildes. *Arch. Naturegesch. Neunzigster Jahrg.,* Abt. A, I Heft: 29–164.

Index

aardvark, *see Orycteropus afer*
Acinonyx jubatus, 9, 10
adults, age of, 60; in agonistic interactions, 122; antipredator behavior in, 152; and cattle egrets, 163; challenged by nonadults, 171; condition in, 94; counts of 44, 45; drinking by, 101; head sparring by, 172; head tossing by, 138; horn shapes in, 27; identification of, 41; licking by, 103; and lions, 151; observed grazing, 87–8; old, 113, 117; pigmentation in, 31; and rank, 121; ruminating by, 115, 117; young, 123
Aepyceros melampus, 9, 10, 11, 84, 87, 217, 221
age, aging signs and, 199; agonistic encounters and, 125; browsing and, 95; and calf development, 187, 191, 193, 194; of a calving cow; of captives, 179; composition of herd by, 59–60; differences and similar physical type, 48; hair color and, 27, 29; head sparring and, 171; horn development and, 25, 27; and identification, 131; of a pathfinder female, 130; rank and, 121, 168; subdivision by, 42; tooth wear and, 124
aggregates, tendency to form, 137
aggression, against humans, 140–3, 191; against vegetation, 167; against cattle egrets, 163; against vehicles, 142–3; by wounded individuals, 140–1; defensive, 156; extraspecific, 156, 163, 200; open intraspecific, 121; preventive, 152; *see also* combat conflict
aging factors, 60

agonistic behavior, 138–79; aggressive–defensive patterns, 143, 149; in bachelor club, 133; the "clinch," 173; confirmatory, 168; elements of, 181, 187; escalation of, 156; extraspecific, 135, 162; gradation of, 172; head-on display, 175; high-intensity, 167, 168–70, 178; horn display, 174; intensity of intraspecific, 167; interspecific, 100; lateral display, 168–9, 174, 175, 178; low-intensity, 121, 122, 123, 124, 168, 175; medium-intensity, 167, 168; nonviolent, 122; in one-sided rank-enforcement, 174; placing head over/under other individual, 175, 176; reluctance to give ground, 168, 174, 175; in sire selection, 33, 180; size display, 168–9; slow movement, 174; and wallowing, 187; *see also* particular agonistic behaviors; aggression, antipredator, avoidance, bar-biting, caution, charging, combat, conflict, defensive, head sparring, head tossing, probing, scrutinizing, threat, etc.
aircraft, observations from, *see* observations, responses to, 139–40
Akeley, C. E., 53
Alcelaphini, 219
Alcelaphus, 216; *buselaphus*, 1, 195; *b. cokii*, 9, 143; *lichtensteini*, 84, 87, 141, 144, 221, 226, 227, 228
alert behavior, 100, 112, 131, 134, 137, 139, 150, 151, 152, 154, 155, 158, 159, 160, 190, 198
alert posture, 73, 112, 130, 131, 135, 138, 139, 151, 152, 154, 155, 159, 161, 162,

245

alert posture (*cont.*)
178, 179, 180, 182, 185, 186, 198, 199,
200; and wind gusts, 179
Algeria, Oued Bou Sellam, 20
allelomimetic behavior, 57, 179, 197–9
anatomy, comparative, 25
ancestral type, 31, 32, 34
Angola, 4
Ansell, W. F. H., 48, 151, 184
antbear; *see Orycteropus afer*
Antidorcas marsupialis, 63
Antilocapridae, 217
antipredator behavior, 148, 150, 151, 152,
158, 225; high-intensity, 152; low-intensity, 152
Aonyx capensis, 10
Ardeola ibis; *see Bubulcus ibis*
artiodactyls, 13, 15
Asia, 15, 222
Atilax paludinosus, 10
aurochs; *see Bos primigenius*
Australia, 118, 219, 222, 223; water buffaloes in, 222
avoidance behavior, 159

baboon, *see Papio*
bachelor groups/clubs, 131–7; cohesion in,
62; combat deescalation in, 173–4; competition in, 167; defensive formations
in, 222; females with, 61; with herds,
51, 62; hierarchy in, 123; and lions, 161;
movements of, 131, 133, 134; and old
males, 117; proportion of all males in,
61; several species compared, 227; size
of, 48, 49; structure of, 133, 134, 135; *see
also* bachelors
bachelors, age and condition of, 131; and
lookout functions, 134; agonistic interactions per individual per hour, 122; condition in, 117; extraspecific contacts of,
136; flehmen in, 181; as lion prey, 147,
200; other-interested behavior in, 199–
200; positions in herd, 75; probing by,
156; separation of, 59; and ruminating,
115, 116; *see also* bachelor groups/clubs
Bainbridge, W. R., 31
banteng, *see Bibos banteng*
bar-biting behavior in captives, 179
basic herd, adherence of males to, 124;
and caretaking of calves, 190; definition
and size of, 47, 48, 137, 224; and "mo-

bile territory," 60; as part of herd, 51,
60, 62, 71, 75, 130; in Serengeti, 52; tendency to associate within, 198; transfer
of males from, 168
basic herds, individuals unaffiliated with,
128; interpenetration of, 60–1, 137; unmingling of, 61, 74
Basilio, A., 54
Bate, D. M., 19, 20
Baudenon, P., 4, 25
behavior, animal, 216; bovid, compared,
216–28; early notes, 1, 3
Belgium, Antwerpen Zoo, 116, 167, 172,
179, 181, 184, 186
Belon, P. (Bellonius), 1
Bibos, 216; *banteng*, 218; *gaurus*, 218
biomass, 85
birth; *see* calving
Bison, 16, 18, 216; *bison*, 18, 179, 218, 226,
227; *bonasus*, 218
Bolton, M., 52
bolus, 117
Bos, 16, 21, 118, 180, 186, 187, 216, 219;
grunniens, 218; *indicus*, 101; *primigenius*,
21; *tarus*, 21, 102, 218, 228
Boselaphini, 16
Boselaphus tragocamelus, 16
boss, 27, 169, 171, 172, 175, 176
Botswana, Chobe Park, 52, 55; Chobe
River, 48, 53, 55
Bourliére, F., 54, 87
Bovidae, 7, 13, 15, 23, 63, 178, 196; African, 15; ancestry, 13, 15; boodont, 15;
boselaphine, 16; domestic, 186; fossil remains of, 13; old, 117
bovids; *see* Bovidae
Bovinae, 13
Bovini, 13, 16, 31, 216, 217, 218, 227
Brooke, Sir Victor, 1, 3, 22, 23
browse, 120; species eaten by *S. c. caffer*
96–7
browsers, 32, 217
browsing, 11, 75, 94–5, 133, 150, 195
Bryden, H. A., 3
Bubalus, 16, 18, 22, 187, 216; *antiquus*, 20;
baini, 20; *bubalis*, 9, 19–20, 22, 118, 181,
198–9, 219, 222, 227, 229, Australian
compared with Busanga *S. c. caffer*,
222–4, camping behavior in, 118, 223;
home range, 223–4; social organization
of, 223; *nilssoni*, 20; *vignardi*, 21

252   Index

kinship, 48; lack of proof of, 127
Klein, R. G., 21
kob; see Kobus
Kobus, 217; defassa/ellipsiprimnus, 9, 10, 11,
    84, 87, 143, 145; k. thomasi, 221; leche,
    10, 84, 148; vardoni, 10, 84, 87, 103, 134,
    149, 152; 221
kongoni; see Alcelaphus buselaphus cokii
Kruuk, H., 225
kudu, lesser, see Tragelaphus imberdis;
    greater, see Tragelaphus strepsiceros

Lamprey, H. F., 94
laufschlag, in African bovids, 221
leadership, 49, 57, 59, 61–2, 123, 129, 130,
    134, 136, 137, 155–6
Leakey, R. E. F., 18
lechwe; see Kobus leche
leopard; see Panthera pardus
Leptobos, 16; syrticus, 16, 17
Lepus capensis, 10
Leuthold, W., 4
Libya, Gebel Zelten, 15; Sahabi, 16
licking, 103–4, 186
licks, salt, 64, 104
lion; see Panthera leo
Littlejohn, K. G., 29
Livingstone, D., 3, 50, 53
Logsdon, S. H., 133
Lonnberg, E., 20
Lorenz, K., 124
Loxodonta africana, 9, 10, 11, 32, 84, 87,
    100, 103, 115, 143, 161, 162, 163; physi-
    cal contact with S. c. caffer, 162
Ludbrook, J. V., 191
Lutra maculicollis, 10
Lycaon pictus, 9, 10, 163
Lydekker, R., 3, 23, 25, 31
lying down, 112, 115, 118, 134, 135, 142,
    155, 170, 171, 175, 178, 179, 181, 185,
    192, 195, 196, 200

McBurney, C. B. M., 32
Maclatchy, A. R., 25
Malbrant, R., 4, 25
males, adult, dominance of, 121; aggres-
    sion against humans, 140–1; aggression
    against lions, 160–1; aggression by
    nonadult, 141–2; aggressive behavior
    against vegetation by, 167; bachelor,
    131–7; in basic herds, 41, 125–9, 158;
    copulatory behavior in, 182; defecating

by adult, 120; in defensive arrays, 222;
    displacement activity in, 186; flehmen
    in, 181; herd, 123, 160, 181; herd de-
    mand for, 170–1; hierarchy in, 122–5;
    high-intensity combat between, 168–70;
    isolated, 135; juvenile, leaving mothers,
    194; in limited combat, 172; in lion kills,
    145, 147, 148, 151; in low-intensity in-
    teractions, 175; 176; old, 27, 117, 124,
    145, 160, 199; other-interested behavior
    in, 199–200; rank enforcing between,
    168; role of, in basic herds, 47; selection
    of, for reproduction, 33, 180; subadult,
    making way for females, 46; top, 128,
    129, 133, 134, 135, 136, 139, 175, 177,
    198; unexplained agonistic behavior in,
    179; urinating in, 118; wallowing in,
    176–7, 228
malnutrition, 32, 94, 117
mating; see sexual behavior
migration in S. c. caffer, 33, 63, 143
Miocene, 13, 15, 16
Mitchell, B. L., 79, 144
Miotragocerus, 16
Mloszewski, M. J., 190, 191
Monard, A., 4
mongoose, 10; see also Mungos, Atilax
monkey; see Cercopithecus aethiops (vervet)
Morocco, 1
mortality, 117; from combat, 59
mothers, 31, 168, 170, 171, 186, 187, 188,
    190, 191, 192, 193, 194, 199, 203; dis-
    criminatory behavior, 187; identification
    by calf, 187
Mozambique, 10; Pungwe river, 55; Save
    river, 53, 55
Mungos mungo, 10, 11
musk ox; see Ovibos moschatus

Nigeria, Cross river, Akparabong, 29;
    Yankeri Game Reserve, 4, 25, 54
nilgai; see Boselaphus tragocamelus
Nilsson, E., 20
nonadults, 113, 117, 130, 148, 149, 155,
    163, 190, 193, 199; imitative behavior in,
    193
North Africa, 20, 21, 32; Central Saharan
    Massif, 19; Maghreb, 15; Mediterranean
    coast, 15; Mediterranean littoral, 21;
    Nile Valley, 19, 21; Sahara, 21, 32, 33
Northern Rhodesia; see Zambia
nyala; see Tragelaphus angasi